14426

THE AUTOMATED CITIZEN

SOCIAL AND POLITICAL IMPACT OF INTERACTIVE BROADCASTING

P. Pergler

Occasional Paper No. 14

The Institute for Research on Public Policy
L'Institut de recherches politiques

Montreal

Printed in Canada

ISBN 0 920380 52 2

Legal Deposit Second Quarter
Bibliothèque national du Québec

The Institute for Research on Public Policy
L'Institut de recherches politiques
2149, rue Mackay
Montréal, Québec
H3G 2J2

Founded in 1972, The Institute for Research on Public Policy is a national organization whose independence and autonomy are ensured by the revenues of an endowment fund, which is supported by the federal and provincial governments and by the private sector. In addition, the Institute receives grants and contracts from governments, corporations, and foundations to carry out specific research projects.

The *raison d'être* of the Institute is threefold:

— To act as a catalyst within the national community by helping to facilitate informed public debate on issues of major public interest

— To stimulate participation by all segments of the national community in the process that leads to public policy making

— To find practical solutions to important public policy problems, thus aiding in the development of sound public policies

The Institute is governed by a Board of Directors, which is the decision-making body, and a Council of Trustees, which advises the board on matters related to the research direction of the Institute. Day-to-day administration of the Institute's policies, programmes, and staff is the responsibility of the president.

The Institute operates in a decentralized way, employing researchers located across Canada. This ensures that research undertaken will include contributions from all regions of the country.

Wherever possible, the Institute will try to promote public understanding of, and discussion on, issues of national importance, whether they be controversial or not. It will publish its research findings with clarity and impartiality. Conclusions and recommendations in the Institute's publications are solely those of the author, and should not be attributed to the Board of Directors, Council of Trustees, or contributors to the Institute.

The president bears final responsibility for the decision to publish a manuscript under the Institute's imprint. In reaching this decision, he is advised on the accuracy and objectivity of a manuscript by both Institute staff and outside reviewers. Publication of a manuscript signifies that it is deemed to be a competent treatment of a subject worthy of public consideration.

Publications of the Institute are published in the language of the author, along with an exectuive summary in both of Canada's official languages.

PREFACE

"It's a small world" is a familiar phrase in a global society linked from end to end by undersea telephone cables, atmospheric radio wave transmitters and television satellites hovering miles above the earth's surface. The technological explosion of the last century is nowhere more evident than in the communications field, where technological developments have indeed made the world "smaller."

It seems paradoxical, then, that while the means of communication are so readily available, communication gaps are as prevalent today as ever--gaps between generations, sexes and cultures; between educators and the less educated; and between citizens and government decision makers, to name a few. Although technology may be a tool for solving communication problems, it can provide only a pathway for the transmission of information: how, by whom, and to what end these pathways are used is determined by those who control them.

Electronic broadcasting has assumed a major role in modern society as radio and television have become the prism through which many people perceive the outside world, from situation-comedy shows to international news and public affairs programming. By their nature, these electronic messages are selective, being manufactured and edited for general consumption by the broadcasting industry, the actors themselves, corporate advertisers, and even political parties. As such, this communication medium provides ample opportunity for the manipulation, unintentional or purposeful, of the general public.

This report deals with the development of interactive broadcasting. "Incasting" is a new technology that permits television viewers to communicate instantly and directly with the broadcasting source. Its potential applications are many and varied, and it is already being used to provide consumer services, educational programming, and political polling and feedback mechanisms, as well as an entertainment function. The author's question is: "Utopia or 1984?" Will such a technology succeed in bridging communication gaps, or will it offer an opportunity for developing vastly more sophisticated manipulative techniques? Interactive broadcasting will soon be introduced to Canada on a commercial scale. The author's message is that steps must be taken now to ensure that it is wisely used. Only that way can we determine whether it will provide a boon or a menace.

Gordon Robertson
President
August 1980

PRÉFACE

"Le monde est petit"; voilà une expression courante dans une société mondiale reliée de bout en bout par des câbles de téléphone sous-marins, des émetteurs-radio et des satellites de télévision suspendus à des milles au-dessus de l'atmosphère. L'explosion technologique de ce siècle n'est nulle part plus évidente que dans le domaine des communications, domaine où les développements technologiques ont effectivement "rapetissé" le monde.

Il semble donc paradoxal que, tandis que les moyens de communications sont si accessibles, les fossés dans les communications soient toujours aussi profonds--fossés entre les générations, les sexes et les cultures; entre les éducateurs et les moins instruits; et entre les citoyens et les décisionnaires du gouvernement, pour n'en nommer que quelques-uns. Même si la technologie peut servir d'outil pour la solution des problèmes de communication, elle ne peut quand même que fournir un fil conducteur pour la transmission de l'information : ce sont ceux qui les ont en main qui décident comment, par qui et à quelles fins ces fils conducteurs seront utilisés.

La diffusion électronique a acquis un rôle de premier plan dans la société contemporaine à mesure que la radio et la télévision devenaient le prisme à travers lequel de nombreuses personnes voyaient le monde extérieur, que ce soient des téléromans, des actualités ou des émissions d'affaires publiques. Ces messages électroniques sont sélectifs de par leur nature même : ils sont fabriqués et composés pour une consommation générale par l'industrie de la radiodiffusion, les comédiens eux-mêmes, les commanditaires privés et même les partis politiques. À ce titre, ce moyen de communication offre nombre d'occasions de manipuler le public, délibérément ou non.

Ce rapport traite de la mise au point de la radiodiffusion interactive. Il s'agit d'une nouvelle technologie qui permet au téléspectateur de communiquer instantanément et directement avec la source de diffusion. Ses utilisations éventuelles sont nombreuses et diverses; elle fournit déjà des services aux consommateurs, des émissions éducatives, des techniques de sondage et de rétroaction politiques et du divertissement. L'auteur demande : "Utopie ou 1984?" Une telle technologie réussira-t-elle à combler le fossé de communications ou permettra-t-elle de mettre au point des techniques de manipulation encore plus avancées? La radiodiffusion interactive se fera bientôt au Canada à l'échelle commerciale. L'auteur soutient que l'on doit adopter dès maintenant des mesures propres à en assurer une sage utilisation. Ce n'est qu'en ce faisant que nous pourrons déterminer s'il s'agira d'un bienfait ou d'une menace.

Le président,
Gordon Robertson
Août 1980

ACKNOWLEDGEMENTS

The present text, in a way that is increasingly typical, results from contributions of many participants, the author being just one of them. The work started by encouragement and advice from Zavis Zeman of The Institute for Research on Public Policy and has used the bibliographic research of Russell Wilkins, also of The Institute. Richard Nichols did part of the background research, and among the experts who gave professional consultations were Peter Parkinson and Gordon Thompson of Bell Northern Research, Michel Rochon and Jan Gecsei of the University of Montreal, F.J. Fletcher, York University, and André Laroque from the Government of Quebec. Several companies who sell the described technology have made available the relevant documentation.

The first rough draft was oriented heavily to the general question of the problem of the identification of the public's attitudes by governments. Dr. David Hoffman, Director of the Futures Studies Program and project leader of this study, started by giving the original manuscript a narrower focus and ended by virtually rewriting several parts of the document, thus becoming more a collaborator than editor.

Nevertheless, the opinions, as well as the omissions and errors, in the text are those of the author.

TABLE OF CONTENTS

EXECUTIVE SUMMARY

The average Canadian, in common with citizens of other developed nations, enjoys the use of a highly complex, home-centred information system. Radio, television, telephone, and mail combine with a wide range of goods, services, and information; enough to allow the average citizen to stay within his home, should he choose to do so, and remain an informed and entertained citizen, possessed of the material goods necessary to his existence.

The technological revolution that has allowed this continues. The western and westernized countries are currently conducting experiments in wired communities. They are proving to be commercial successes. The wired communities are an advance over the present television-radio-telephone system in that all information, incoming to the home and outgoing, is centralized in one device, essentially a television set altered and expanded to provide more services than entertainment and information. This alteration of the television set forces the invention of a new word, *incasting*. If broadcasting is the term applied to communications to the citizen from the radio or television source, then *incasting* may be used to describe upstream communication from the citizen to the radio or television source, government, or others who affect his life. Incasting involves an interaction between the citizen and those who seek to inform him, educate him, entertain him, regulate him, or provide him with goods and services.

The governments of countries in which these changes are taking place have not attempted, by and large, to assist or regulate the new incasting systems. In Canada, there are no proposals in the public domain to regulate or encourage the experimental systems. This lack of action could result in the development of incasting systems in Canada along lines unfavourable to the public interest. It would be a matter of luck, given a lack of public input, control, or assistance, whether incasting systems improved the quality of life that citizens enjoy in Canada.

INTRODUCTION

There are two standard scenarios offered by science fiction writers and communications theorists with respect to how the two-way television set will affect the citizen. Essentially, either the citizen will control the television set, and influence the world around him through its incasting capability, or the television set will condition and control the citizen, and through it, those in power will reach and influence the citizen. In the latter, the citizen would be agreeably controlled. He would be an automated citizen. George Orwell's *1984* is a fictional exploration of rebellion against automated citizenship and television control, and Ray Bradbury's *Fahrenheit 451* is, perhaps, the first serious attempt in fiction to deal with the possibility of incasting. In the few short years since Orwell and Bradbury wrote their novels, incasting has become a reality.

CHAPTER ONE

There are, of course, more traditional methods of incasting. The citizen has always been able to reply or respond to messages broadcast to him by radio or television stations, either by telephoning the station, or by writing the station a letter. Telephone polls or surveys and mail questionnaires are other methods often used to get the listener's opinion, but studies have shown these to be clumsy and incomplete methods of estimating the opinion of the total audience. They are awkward tools.

Home information systems currently being developed and in experimental use will allow more sophisticated and automatic polls, surveys, and sales. For example, if a television network broadcasting to a wired community wanted to have a sample of audience reaction to a new programme, it could automatically sample its entire wired audience, detailing the percentage that turned the programme off after a few seconds, the percentage that watched the entire show, and any other viewing characteristic that the television executive or advertiser wished to have. The technology exists to provide this sampling with or without the viewer's consent or knowledge. Further, if the characteristics of the families viewing are known, then the sampling in the wired community could be selective, should the advertising executive be interested only in the response of various segments of the community, for example, the rich, or the elderly, or families with children still at home.

Apart from automated polls or surveys, interactive broadcasting could, and in the current experimental wired communities does, perform a variety of tasks and functions. Of immediate interest to the viewer is an expanded choice of programmes, an expansion of the pay-television or cable-television concept. Other functions capable of being performed by the incasting television are alarm systems for fire, home security, or medical alerts, and the remote sensing of home utility meters. Centrally stored information, from advertisers or computers, can also be made available to the home screen. The interactive television, wired to communications lines, can electronically transfer funds, pay creditors, and allow the viewer to purchase goods advertised on television. Again, video games, through interactive broadcasting, can link up competitors in distant locations, or allow the player to play against centrally stored programmes. Other and similar incasting functions are, potentially, specific channels for public voting, and the counting of votes at parliamentary assemblies by electronic means.

CHAPTER TWO

There are more wired communities and examples of interactive broadcasting at present than are generally known. Perhaps the outstanding example is QUBE, a 30-channel, cable television system operating in Columbus, Ohio. QUBE allows pay television for selected premium programmes, free viewing of many public programmes, and sponsored programmes. And the viewer may register his opinion of what he is watching through a push-button console. Questions asked through the home video screen about consumer and public issues may also be immediately responded to through the five-button choice. QUBE is owned and operated by the giant American entertainment conglomerate, Warner Communications, Inc., and is generally considered by viewers and Warner to be a success.

Votaphone is a voting and polling mechanism being developed by Bell Northern Research of Canada in its Ottawa laboratories. It is connected to the home telephone but may be used independently of the telephone. Votaphone could be developed to serve such functions as audience size estimation, automatic meter reading, pay-television billing, and alarm systems reporting, but it has not yet been so developed.

Vista is another Bell Canada experiment. Basically a videotext system, it can make available to a home screen a vast amount of information usually accessible only through libraries or computers. The 1979-80 experiment involves 25 users' terminals making use of 2,000 pages of text, in French and English, in seven colours; the text consists of current information on news, weather, sports, classified advertising, consumer services, and entertainment reading. It is, in essence, an electronic newspaper. During 1979, a new improvement in Canadian videotext was being developed in conjunction with Vista. Telidon, developed by the federal Department of Communications, may prove superior to other videotext systems. A Telidon test in 1979 involved 120 homes in Alberta and Ontario.

Other incasting attempts or systems have been tried in Canada and the United States. As a general statement, however, none approach the HI-OVIS system introduced in 1978 in western Japan. HI-OVIS is an interactive television system using optical fibre. The first test involved 158 homes. Each home received a colour television receiver, a terminal adapter, a camera, a microphone, and a light receiver-transmitter. By using push buttons, the viewer could choose from any of 29 programme channels. He could also ask questions and receive instantaneous replies. This test situation is reported to be working well. The Japanese optimistically hope for a nation-wide system availability by 1985.

Other interactive or two-way, private enterprise television systems are in operation in The Woodlands near Houston, Texas, in Fort Lauderdale, Florida, and Menlo Park, in California. Generally they add a response element to cable television systems. Television-connected and incasted fire and medical alert systems have resulted, in some cases, in reductions of insurance premiums paid by members of these wired communities.

Speculation

The social effects of incasting systems on their communities are a matter of speculation. There have been no studies and the wired communities are themselves too recent to manifest easily observable changes. Some newspaper articles on the effect of QUBE on Columbus, Ohio suggest that QUBE viewers are not being manipulated by their involvement in the QUBE system, but that their opinions and attitudes are often in advance of what QUBE is prepared to offer them. For example, a high percentage of QUBE viewers who selected Channel P-10, the soft-core porno channel, complained of excessive editing.

At present, interpretation of viewer incasting in QUBE and other systems presents considerable problems. For example, as television seems to be going through a radio-like phase from centre of attention to background, with an increasing percentage of viewers keeping their sets on all day whether anyone is watching or not, how reliable would automatic, station-controlled samples of viewer preferences be? Moreover, since there are as yet only a few experimental wired communities and these communities know themselves to be experimental, and, perhaps, think themselves "special" or "temporary," the applicability of this experimental data to regularized programming is questionable.

CHAPTER THREE

It seems likely that electronic incasting will soon be widely available to citizens of western and westernized countries. Private enterprise appears at present to be leading the introduction of these systems as part of the entertainment industry. However, as the present means of obtaining citizen input on social and political questions is widely regarded as inadequate, it seems equally likely that the systems will be developed to allow social and political polling as well. Current and extensive calls for direct democracy could well be fulfilled through incasting television transceivers. Members of political assemblies could receive constantly updated polls of their constituents' attitudes. The public service, at present widely thought to be out of touch with the public's needs and general attitudes, could profit from incasting information. The possibility of citizen feedback reaching directly the political representative or public servant could well spark greater citizen interest in social issues.

Complications

Some unique features of the Canadian situation may affect the introduction of interactive broadcasting.

1. The federal-provincial conflict over the responsibility for communications may inhibit development, and regulation of incasting systems.

2. Our great size, narrow population zone, cultural differences, and short history inhibit communication; this might be partially overcome by a rational incasting system.

3. The traditionally strong position of the public service in Canada makes communications to government more necessary than in other countries. Incasting could contribute.

4. The overwhelming cultural and economic influence of the United States in Canada might be lessened by incasting systems.

5. In Canada, a conflict exists between the carriers of transmitted programmes and the operators responsible for content. This leaves the home entertainment industry in a situation of change and uncertainty. The introduction of incasting systems would therefore take place in a milieu of considerable flux.

GENERAL CONCLUSION

Business interests, without caring much about social incasting, will probably soon introduce to Canada technologies and organizations that will make social incasting possible. Once introduced, social incasting will contribute to the pressure for improvement of overall government-citizen communication. If governments undertake a profound revision of this process of communication, incasting may marginally help. If they do not, incasting will marginally exacerbate the coming conflict.

PRINCIPAL RECOMMENDATIONS

The government should start, as soon and as intensively as possible, to make improvements in the process by which the government (*a*) understands the perceived needs and policy-related attitudes of Canadians; and (*b*) anticipates the reactions of Canadians to government actions. In this context, which is crucial both for the incasting and for much broader concerns, the government should move to identify and facilitate the main role of interactive broadcasting.

The government should prepare regulations for interactive broadcasting while not overestimating how much any regulation can influence interactive broadcasting in Canada. It is desirable to start the work soon and continue as much as possible without pressure, to try to achieve at least communication between actors, before the positions become petrified and the conflict starts in full.

Government funds should be directed to the development of Canadian interactive broadcasting according to economic criteria (for example, employment, profits, learning process).

ABRÉGÉ

Le Canadien moyen, à l'instar des citoyens des autres pays développés, jouit de l'usage d'un ensemble de moyens d'information domestiques très complexes. La radio, la télévision, le téléphone et la poste s'unissent à une vaste gamme de biens, de services et d'informations, à tel point que le citoyen ordinaire peut, si tel est son désir, demeurer à la maison en restant informé et diverti et n'étant privé d'aucun des biens matériels nécessaires à son existence.

La révolution technologique responsable de cette situation se poursuit. Les pays occidentaux et occidentalisés font actuellement l'expérience de communautés câblées qui s'avèrent un succès commercial. Elles marquent un progrès par rapport au réseau télévision-radio-téléphone actuel en ce que toute l'information qui entre à la maison ou en sort est concentrée dans un seul appareil : il s'agit essentiellement d'un téléviseur modifié et amélioré de façon à procurer des services autres que les seuls divertissement et information. Cette modification du téléviseur nous permet maintenant de parler de *radiodiffusion interactive*. Si la radiodiffusion désigne la communication dirigée vers le citoyen depuis la radio ou la télévision, la radiodiffusion interactive désigne la communication dialoguée, du citoyen vers la source de télédiffusion ou radiodiffusion, vers le gouvernement ou vers d'autres qui affectent sa vie. La radiodiffusion interactive comporte une interaction entre le citoyen et ceux qui cherchent à l'informer, à l'éduquer, à le divertir, à le réglementer ou à lui procurer biens et services.

Les gouvernements qui assistent à ces changements n'ont pas, somme toute, tenté de favoriser ou de réglementer les nouveaux modes de radiodiffusion interactive. Au Canada, on ne connaît aucun projet d'ordre public pour l'encouragement ou la réglementation de ces systèmes expérimentaux. Cet immobilisme pourrait entraîner un développement des moyens de radiodiffusion interactive au Canada qui soit contraire à l'intérêt public. Sans participation, aide ou contrôles publics, nous pourrions nous compter chanceux de les voir améliorer la qualité de vie des Canadiens.

INTRODUCTION

Les auteurs de science-fiction et les théoriciens des communications présentent habituellement deux scénarios quant aux effets de la télévision bidirectionnelle sur le citoyen. En quelques mots, ou bien le citoyen aura la maîtrise de son téléviseur et pourra influencer son milieu grâce à son pouvoir interactif, ou bien le téléviseur conditionnera et dirigera le citoyen et, de ce fait, les détenteurs du pouvoir l'atteindront et l'influenceront. Le citoyen consentirait en l'occurrence au contrôle. Il deviendrait un citoyen automatisé. *1984*, de George Orwell, est un compte rendu romancé de la révolte contre la vie automatisée et l'emprise de la télévision, tandis que *Fahrenheit 451* constitue peut-être la première tentative sérieuse d'explorer, dans le roman, la possibilité de la radiodiffusion interactive. Au cours des quelques années écoulées depuis la publication des romans d'Orwell et de Bradbury, la radiodiffusion interactive est devenue une réalité.

CHAPITRE PREMIER

Il existe évidemment des méthodes de radiodiffusion interactive plus classiques. Le citoyen a toujours eu la possibilité, que ce soit par le téléphone ou par la poste, de répondre ou de réagir aux messages diffusés par les stations de télévision. Les enquêtes ou sondages téléphoniques et les questionnaires postaux sont d'autres façons habituelles d'obtenir l'opinion de l'auditeur, mais des études ont révélé qu'il s'agit là de moyens grossiers et incomplets de juger de l'opinion de l'ensemble de l'auditoire. Ce sont des outils peu maniables.

Les moyens d'information domestiques que l'on est actuellement à mettre au point et à l'essai permettront des sondages, enquêtes et ventes plus automatiques et raffinés. Si, par exemple, une chaîne de télévision diffusant à une communauté câblée voulait échantillonner la réaction de l'auditoire à une nouvelle émission, elle pourrait le faire automatiquement et ce pour l'ensemble de la population câblée, tout en obtenant des renseignements sur la proportion des gens qui avaient coupé l'émission après quelques secondes ou qui avaient regardé l'émission entière ou sur toute autre caractéristique que la direction de la station ou le commanditaire souhaite connaître. La technologie permettant cet échantillonnage, avec ou sans l'accord du spectateur, existe déjà. Si on connaît, en outre, les caractéristiques des familles spectatrices, l'échantillonnage de la communauté câblée peut être sélectif au cas où le commanditaire s'intéresserait avant tout à la réaction de divers segments de la communauté, tels que les riches ou les personnes âgées ou encore des familles avec des enfants à la maison.

La radiodiffusion interactive (et cela existe déjà dans les communautés câblées), en plus de permettre des enquêtes ou sondages automatisés, peut remplir diverses fonctions et effectuer diverses tâches. Le spectateur s'intéressera en premier lieu au choix d'émissions plus étendu et à l'extension des concepts de télévision payante et de télévision par câble. La télévision interactive peut aussi remplir d'autres fonctions : système d'alarme en cas de feu, de vol ou d'urgence médicale, ou lecture à distance des compteurs domestiques. On peut aussi diffuser vers les maisons des informations fournies par les annonceurs ou les ordinateurs et stockées dans des zones de mémoire centralisées. La télévision interactive branchée à des réseaux de communication peut virer des fonds électroniquement, rembourser les créanciers et permettre au spectateur d'acheter des produits annoncés à la télévision. On peut aussi permettre aux amateurs de jeux électroniques de se mesurer à des adversaires éloignés ou de jouer contre des programmes stockés dans une zone de mémoire centrale. On peut aussi penser à des canaux destinés à des votes publics et au dépouillement du scrutin lors d'assemblées parlementaires.

CHAPITRE DEUX

Il existe davantage de communautés câblées et d'exemples de télévision interactive qu'on le croit généralement. L'exemple le plus remarquable est probablement QUBE, un réseau de télévision par câble de 30 canaux à Columbus, en Ohio. QUBE permet de payer pour certaines émissions de choix et de visionner gratuitement plusieurs émissions publiques ou commanditées. Le spectateur peut faire connaître son opinion sur l'émission qu'il regarde au moyen d'une console à boutons-poussoirs. Les cinq boutons permettent aussi de répondre instantanément à des questions relatives à la consommation et aux affaires publiques transmises sur l'écran. Le gigantesque conglomérat de divertissement américain, Warner Communications Inc., est propriétaire-exploitant de QUBE. Warner et ses spectateurs sont d'avis qu'il s'agit là d'un succès.

Bell Northern Research of Canada est à mettre au point, dans ses laboratoires d'Ottawa, un mode de scrutin et de sondage nommé *Votaphone*. Le système est branché sur l'appareil de téléphone domestique mais on peut s'en servir indépendamment. On pourrait le développer à des fins d'estimation de la taille de l'auditoire, de lecture automatique des compteurs, de facturation de la télévision payante et de surveillance des systèmes d'alarme, mais cela n'est pas encore fait.

Vista est une autre expérience de Bell Canada. Il s'agit essentiellement d'un système vidéotex qui peut transmettre sur l'écran domestique une somme énorme d'informations habituellement disponible uniquement dans des bibliothèques ou des ordinateurs. L'expérience tentée en 1979-1980 comprend les terminaux de 25 utilisateurs et 2 000 pages de texte en français et en anglais ainsi qu'en sept couleurs; il s'agit de renseignements sur l'actualité, la météo, les sports, les services aux consommateurs, d'annonces classées et de textes de divertissement. C'est en somme un journal électronique. Au cours de 1979, on était à améliorer le vidéotex canadien de concert avec Vista. Telidon, mis au point par le ministère fédéral des Communications s'avérera probablement supérieur aux autres systèmes vidéotex. Un essai du Telidon en 1979 comportait 120 foyers en Alberta et en Ontario.

On a fait l'essai, tant au Canada qu'aux États-Unis, d'autres systèmes de radiodiffusion interactive. Dans l'ensemble, cependant, aucun d'eux ne s'approche du système HI-OVIS introduit en 1978 dans l'ouest du Japon. HI-OVIS est un système de télévision interactive utilisant la fibre optique. Le premier essai comprenait 158 foyers. Chacun reçut un télécouleur, un raccord de terminal, une caméra, un microphone et un petit récepteur-émetteur. En appuyant sur les boutons-poussoirs, le spectateur pouvait choisir entre 29 canaux. Il pouvait aussi poser des questions et obtenir des réponses immédiates. On rapporte que cet essai se déroule bien. Les Japonais espèrent, avec optimisme, que le système sera disponible de par le pays d'ici 1985.

On retrouve d'autres systèmes bidirectionnels ou interactifs exploités par l'entreprise privée à The Woodlands, près de Houston, au Texas, à Fort Lauderdale, en Floride, et à Menlo Park, en Californie. On y ajoute habituellement un mécanisme répondeur au système de télévision par câble. Des systèmes d'alarme pour le feu et les urgences médicales branchés sur la télévision interactive ont parfois occasionné une baisse des primes d'assurance des utilisateurs.

Quelques spéculations

Les effets sociaux des systèmes de radiodiffusion interactive sont matière à spéculation. Il n'existe aucune étude, et les communautés câblées sont d'origine trop récente pour que l'on puisse y déceler facilement quelque effet que ce soit. Certains articles de journaux relatifs à l'effet de QUBE à Columbus, Ohio, laissent entendre que la participation des abonnés de QUBE à ce système n'entraîne pas leur manipulation; leurs attitudes et opinions seraient plutôt souvent plus évoluées que ce que QUBE est en mesure de leur présenter. Par exemple, un pourcentage élevé des spectateurs choisissant le canal P-10, qui offre de la pornographie suggestive, se sont plaints d'une censure excessive.

À l'heure actuelle, l'interprétation de l'interaction entre le spectateur et QUBE ou d'autres systèmes est parsemée d'embûches. Puisque, par exemple, la télévision semble être en voie de passer, tout comme naguère la radio, du centre d'attention à un bruit de fond, alors qu'un pourcentage croissant de spectateurs la laissent allumée toute la journée, que quelqu'un la regarde ou non, dans quelle mesure pourrait-on se fier à des échantillonnages des choix des spectateurs réalisés automatiquement à partir des stations? En outre, comme il n'existe, à ce jour, que quelques communautés câblées expérimentales et qu'elles se savent telles et se croient peut-être "spéciales" ou "temporaires", on peut douter de la pertinence de ces données expérimentales pour la programmation ordinaire.

CHAPITRE TROIS

Il semble probable que les citoyens des pays occidentaux et occidentalisés jouiront bientôt d'un accès facile à la radiodiffusion interactive électronique. L'entreprise privée semble, à ce jour, être la première à introduire ces systèmes comme composantes de l'industrie du spectacle. Par contre, puisqu'on considère comme généralement inadéquats les moyens d'obtenir le point de vue des citoyens sur des questions sociales et politiques, il est probable que l'on mettra au point des moyens de sondages social et politique. Les appels actuels généralisés à une démocratie directe pourraient bien avoir leur réponse grâce à des émetteurs-récepteurs combinés interactifs. Les membres des assemblées politiques pourraient être constamment tenus au courant des attitudes de leurs commettants. La fonction publique, que beaucoup considèrent comme détachée des besoins et attitudes du public, pourrait profiter de la radiodiffusion interactive. La possibilité que la réaction des citoyens atteigne directement le représentant politique ou le fonctionnaire pourrait bien éveiller davantage l'intérêt du citoyen aux questions sociales.

Les complications

Certains traits propres à la situation canadienne pourraient influer sur l'introduction de la télévision interactive.

1. Le contentieux fédéral-provincial relatif à la responsabilité en matière de communications pourrait nuire au développement et à la réglementation des systèmes de radiodiffusion interactive.

2. Notre vaste étendue, notre petite population, nos différences culturelles et notre courte histoire nuisent à la communication; un système rationnel de radiodiffusion interactive pourrait aider à surmonter ces obstacles.

3. La position de force traditionnelle de la fonction publique au Canada rend les communications avec le gouvernement plus nécessaires que dans d'autres pays. La radiodiffusion interactive pourrait y aider.

4. La radiodiffusion interactive pourrait pallier l'écrasante influence culturelle et économique des États-Unis sur le Canada.

5. Au Canada, il existe un conflit entre les diffuseurs des programmes transmis et les exploitants responsables du contenu. L'industrie du divertissement domestique se retrouve, de ce fait, dans une situation de changement et d'incertitude. L'introduction de systèmes de radiodiffusion interactive se produirait donc dans un contexte passablement changeant.

CONCLUSION GÉNÉRALE

Le milieu commercial, qui se soucie peu de la radiodiffusion interactive sociale, introduira probablement sous peu au Canada les technologies qui la rendront possible. Une fois introduite, la radiodiffusion interactive sociale accroîtra probablement les pressions en vue d'améliorer l'ensemble des communications entre le gouvernement et les citoyens. Si les gouverrements révisent en profondeur le processus de communication, la radiodiffusion interactive aidera quelque peu. Sinon, elle exacerbera quelque peu le conflit à venir.

LES PRINCIPALES RECOMMANDATIONS

Le gouvernement devrait, aussitôt et aussi intensément que possible, commencer à améliorer le processus qui lui permet de *a*) comprendre les besoins perçus chez les citoyens et leurs attitudes face aux politicues et *b*) prévoir les réactions des Canadiens aux actions gouvernementales. Dans ce contexte, d'importance primordiale tant pour la radiodiffusion interactive que pour d'autres raisons plus englobantes, le gouvernement devrait agir afin d'identifier et favoriser le rôle principal de la diffusion interactive.

Le gouvernement devrait préparer une réglementation relative à la diffusion interactive sans toutefois surestimer l'influence de quelque réglementation que ce soit sur la radiodiffusion interactive au Canada. Il est à souhaiter que le travail s'amorcera sans tarder et qu'il ne se fera pas trop sous pression afin, au moins, de permettre aux acteurs de communiquer avant que leurs positions ne se durcissent et que le conflit n'éclate.

Les argents gouvernementaux devraient être orientés vers le développement de la radiodiffusion interactive canadienne en fonction de critères économiques (tels que l'emploi, les profits, le processus d'apprentissage).

INTRODUCTION

IMAGE ONE

I am sitting in a comfortable chair in front of my big, three-dimensional, colour television set. Any information, entertainment, or message I choose is presented to me immediately. I enjoy the apex of communications development that started with the use of human language and progressed through pictures in the sand, writing, printing in movable type, newspapers, telegraph, telephone, radio, to television. I am a perfectly informed person.

A voice receiver on the television set accepts my instructions. It has replaced the push-buttons of the eighties, and may soon be replaced by sensors of electrical impulses in my brain. I decide what to buy and see immediately the impact on my personal accounting; what game to play with any of my friends, however far away they may live; in which course, at which university I am to be a student. If I wish to call a physician, a plumber, or a team of family advisers to treat an emergency, they are all accessible through my voice-activated information machine.[1]

I am also a full participant in social and political decisions around and about me. Whenever I find it relevant, my voice is heard--as an opinion or as a vote. Without unnecessary time delays, from any place, from any person, the inputs from all concerned citizens enter the decision-making process. The development from Hammurabi and Plato to ....(please fill in the name of your favourite prophet) has come to its full completion. In the socio-political sense, I am a free person.

IMAGE TWO

I sit in the same chair in front of the same television set with the same voice receiver. I can choose any channel I wish and I will always like some of the programmes. Knowing exactly the taste of the population, the Establishment (who, of course, also runs FICN, our Fully Integrated Communication Network) has effectively devised ways of knowing how to please everyone, including themselves. The format and shape fit the viewer's taste, the content fits the Establishment's needs.

There is a qualitative difference with the past: my input capability is nearly saturated by FICN and other sources of information can have only a negligible impact on me. Also, other sources seem less perfect, less tailor-made to my taste, less fun than FICN. Years of conditioning have been successful. The dynamics of communication have come to an end, for there is no motivation as a driving force of change. I am a completely and irrevocably misinformed person.

I am free to give any instructions to my television set and they will always be obeyed. Effective procedures helped to optimize the expenses for and the orientation of programmes in order to achieve a desirable statistical distribution of reactions. The beginning of this scientific approach has been a creation of perceived "need to buy" by mass media advertisement and the use of media in pre-election and other political campaigns. The Establishment knows how to get the right answers as a result of subjectively free decisions. It controls the information to the viewer, timing and wording of questions, elaboration and interpretation of conclusions, application and dissemination of results. I am (in the stochastic sense and from the socio-political point of view) completely and agreeably controlled. I do not even object, I am an *automated citizen.*

Of course we are familiar with both of these images. They are the standard components of several science-fiction models and, indirectly, of our occasional late night discussions, if malt Scotch has provoked in us a desire to save the world.

The present text explores the possibility that there is more to these images than dreams. We are concerned with the impact of electronic channels that may transfer the information, especially with the impact of interactive broadcasting, that is, the possibility of a viewer communicating directly with the broadcasting centre.

The developing technology is turning this possibility into reality. A few basic technological principles can now perform a surprisingly broad cluster of functions (Chapter 1), some of which are directly related to our topic. There are already some examples of practical applications (Chapter 2). It is working well and will work better.

Our difficulties start with questions of what it should work for? We seem tacitly to assume that our process of governing needs the technology and the methods that have been identified as coming. But does it? Who believes so? The final chapter is concerned with the governing process as it exists, and with some components that do not but perhaps should exist. Given the nature and character of the existing process, what role, if any, can be expected from the oncoming technologies and methods? Are any conclusions at all possible? Can we even make some recommendations?

If any conclusions are at all possible, they should probably be called "conditonally optimistic." If we are lucky, the new technologies could trigger, or at least facilitate, courses of action that would be generally helpful. If we are less lucky, the same technologies could do significant harm.

The tenor of this study is this: if we drift and do nothing, we shall, in this context, be less lucky.

CHAPTER ONE

PRINCIPLES

As human communication developed, so did the development of communication patterns. Monologue, dialogue, and what we now call conference--a pattern of mutual dialogues in a group--have been with us as long as we can remember (see Table 1). The word "broadcasting" came into use with radio. More generally, it now includes continual or periodical dissemination of information from one centre by any electronic means--word or picture, wireless or cable. A possible extension to the concept of "mass media," which would make the classification generally valid, is, for our purpose, of only marginal importance.

The concept of "*incasting*" is more recent and describes an "upstream capability" of a broadcasting network: the possibility of any listener or viewer transmitting a message to the broadcasting centre. In most cases, the large volume of expected incoming messages makes statistical summarization necessary, which in turn limits the message to only a few bits of information. The present work is concerned mainly with incasting. We begin by summarizing the most relevant technological principles that are used within the incasting pattern (see Table 2). The reader who does not need to be convinced about the technical feasibility and applicability of incasting may wish to jump to Chapter 2.

Of all traditional technologies, mail is still an effective way for answering requests for opinions by the mass media. The answers are not limited in length, timing is flexible, and the respondent has time to think over the questions. With the exception of panels that have already agreed to participate, however, the return ratio is usually low.

The most familiar network for electronic incasting is the telephone. In its original function it transfers audio messages in a dialogue pattern. Its role in interactive broadcasting can be considerably broader and it can use many new technologies (see Table 3). Questionnaire surveys are frequently conducted by phone. This procedure is relatively inexpensive, enjoys a good return ratio (i.e., percentage of contacted persons who answer the questions), and is fast. On the other hand, the sample might be biased: not everybody has a phone; worse, it is not certain who and under what conditions the respondent is really answering. Moreover, the availability and willingness of people to participate changes with time.

Sometimes the first question in a phone poll asks for a list of family members and their ages; then a request is made, for example, for the "second oldest female member of the family" or whatever is necessary to fill the caller's quota. This procedure is practical but presents a conundrum for statistical evaluation. There are also other problems: respondents may be distracted or irritated; or they may distrust the process, its confidentiality and purpose, something which is not always easy to detect by phone.

TABLE 1

COMMUNICATION PATTERNS

Monologue	One-to-self
Dialogue	One-to-one
Conference	Many-to-many
Broadcasting	One-to-many
Incasting	Many-to-one

Of all these communication patterns, our main concern will be with incasting, that is, bringing information from many places to one centre. "Interactive broadcasting" usually combines broadcasting and incasting. For broadcasting and incasting technologies see Table 2.

TABLE 2

COMMUNICATION TECHNOLOGIES

	ACOUSTIC		MATERIAL
*CLASSICAL**	Face-to-face		Mail*
*ELECTRONIC**	CABLE*		HERTZIAN
	COPPER	FIBRE*	
AUDIO*	Telephone*		Radio
ALPHANUMERIC*	Telegraph, teleprint, information processing device and network*, videotex*		
ICONIC: STATIC	Facsimile, screen graphic		
ICONIC: DYNAMIC*	Cabletelevision*		Television

The table describes technologies for the one-to-many pattern, for example, for broadcasting. The asterisk denotes applications for the many-to-one pattern, for example, incasting. The term "radio" is sometimes used in lieu of "hertzian" which relates to communication using electromagnetic waves.

Questionnaire surveys are acquiring a bad image, only partially for valid reasons. The phone survey, work-horse of the profession, is sometimes regarded with outright suspicion. On the other hand, its reputation might improve; indeed, if the quality of the research generally were to rise, there would be always a good niche for simple, fast, and reasonably reliable phone surveys.

There exists another application of the phone for research: radio and television talk shows sometimes ask the listener/viewer to dial manually one of a set of phone numbers indicated during the broadcast and to report his opinion orally. This is a simple, fast, and inexpensive approach. An unknown number of callers, however, never reach the indicated number, as it is busy. The perseverance of some in repeating the call is also a biasing factor in the sampling, but if the phone-in is in response to information published by newspapers, or delivered by mail, this factor can more easily be overcome.

There are other means of overcoming the drawbacks in the phone surveys. Sometimes only a sample of the population is given the phone number and requested to call. It is also possible to equip each member of the sample group with an automatic dialing device. Sometimes, a different number is given for each alternative answer (e.g., for "yes" and for "no") and a simple device just counts calls to each number. Another approach is to initiate the calls from a centre, either manually or by automatic devices. All these methods, however, are incompatible with standard operation of the telephone: while making another phone call, the customer cannot participate in incasting. Even this problem is not in the end insurmountable, however, since with the use of the so-called "test trunk" a central dial-up approach can be used concurrently with a standard house call. Finally, a combined approach is being used increasingly. In this a respondent calls in his phone number, which is usually tape-recorded. The recording takes only a few seconds and is followed, within a reasonable time span, by a return call, so decreasing the time pressure.

TABLE 3

TELEPHONE INCASTING TECHNOLOGIES

Dial-up:	individual	manual automatic
	central	manual automatic test-trunk
	combined	
Non-dial-up:	copper	loop longitudinal
	fibre	

For the difference between the loop and the longitudinal connection see Figure 1.

There are also several ways to perform incasting without dialing, still using the phone network with conductive connections. The oldest way is to equip telephones with an appropriate source, operated, for example, by push-button and to transmit the message via the same loop as the regular phone conversation. In this case, incasting cannot be performed simultaneously with conversation. This difficulty can be overcome with a so-called longitudinal connection, where incasting does not interfere with conversation. (The technology-oriented reader may see Figure 1 for how it is done).

So long as the broadcast comes via cablevision, it seems natural to use the cable for incasting as well. Both copper and fibre connections mostly use a centre-initiated procedure: an automat checks periodically, for example, once every six seconds, all connected customers and records answers. Is the set on? Has any button been pushed within the last six seconds? In this way messages can be both broadcast and incast and the latter recorded centrally.

Alphanumeric systems (see the lower part of Table 2), started long ago by telegraph, go in the direction of electronic mail and integrating information systems. Moreover, their application for incasting is technologically relatively simple, compared with other technological and non-technological problems. There is, however, not much practical experience with the approach.

ALTERNATIVE METHODS OF OPERATION

Table 4 summarizes several dimensions of alternatives that are at present available for incasting. The first alternative concerns *the way in which the reaction of the participant is triggered* (1). It can be an answer to an explicit question ("How do you like this advertisement? Push button 1, if..."), or the participant may have a permanent possibility to communicate whenever he pleases, pushing, for example, button 1 for "You bore me!" Some technologies make both options possible, some only one of them.

When the message arrives at the centre, *each answering participant may or may not be identified* (2). If confidentiality is of importance, it depends upon the question of where the participants are identified. They may be identified in the local telephone switchboard, but their identity is lost before the message reaches the centrum; or in the memory of the central computer; or in some of the standard outputs of this computer, for example, for billing purposes in pay television; or nowhere outside their own telephone loop. Given the usual possibility of leakages, only the last alternative gives a guarantee of anonymity, about as strong as for the content of a normal telephone conversation.

For statistical purposes it is sometimes useful to know selected characteristics or indices of participating households--for example, the income bracket, or type according to age, of members of the household. If the identity of participants is recorded in the central computer memory there are no technical problems, but there may be problems with confidentiality. Compromise solutions are difficult and untested.

It is possible to give the *opportunity to participate*, and to supply the necessary hardware, to every receiver of the broadcasting, or only to selected ones (3). If a selection takes place, the *nature of the sample* may vary: it

FIGURE 1

TWO TECHNOLOGIES FOR INCASTING VIA THE TELEPHONE SYSTEM (see Table 3)

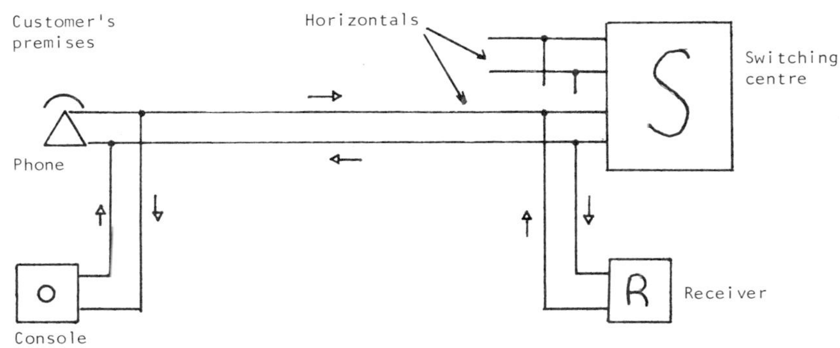

LOOP CONNECTION

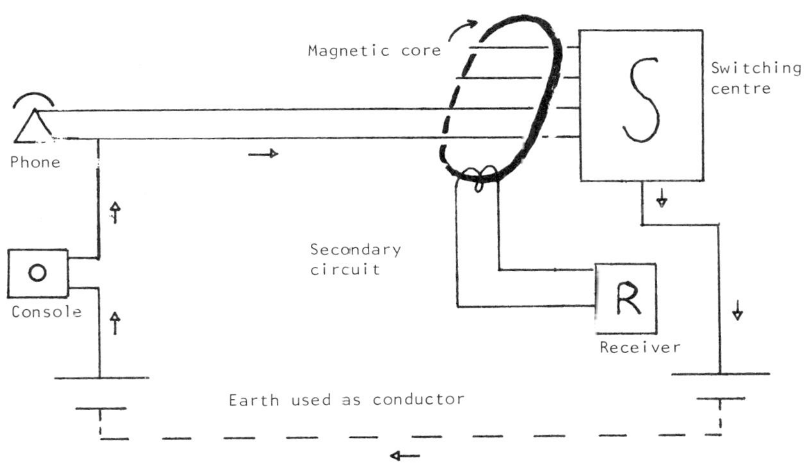

LONGITUDINAL CONNECTION

Note:

The loop connection in the top part of the figure uses a console (i.e., a source and a push-button) and the same loop as used for conversation. This leads to mutual interference.

In the switching centre, the message is identified by a receiver and processed as necessary, for example, summarized, registered, etc. Many pairs of wires run in parallel bundles, called horizontals, and their tapping is impossible. A better way, known as longitudinal connection, is described in the power part of the figure. The circuit uses earth as one of its legs and the regular connectors as the other leg. A magnetic core can be fitted in the switching centre around any pair, any horizontal, or any of their combinations.

Each impulse from a push-button generates an impulse in the secondary circuit, using the core as a transformer. The impulses are received and transmitted without disturbing the usual function of the telephone.

can be purposive (e.g., certain types of households), or random, or self-selected, for example, anyone who wants may participate (4). Selection has obvious importance for possible conclusions that may be drawn from the received reactions. The second limit to statistical interpretation comes from the return ratio which is perceived in incasting as the ratio of the number of reactions to the number of customers who have the technical opportunity to answer and who have switched on the channel.

Originally the survey respondents were scattered over the whole *broadcasting area* (5). More recently the first experiments with so-called "narrowcasting" have brought some success; for instance, a programme about local zoning changes can be sent by cable to the concerned area only. Alternatively, reactions from the concerned area can be isolated and treated separately, while the rest of the audience may watch the programme only as simple information or, more probably, switch to another channel.

TABLE 4

ALTERNATIVES IN INCASTING OPERATIONS

TRIGGERING (1)
Synchronous
Asynchronous

RESPONDENTS (2)
Anonymous
Identified: with indices
without indices

DISTRIBUTION OF CONSOLES (3)
Exhaustive
Selective

SAMPLE (4)
Purposive
Random
Self-selected

DOMAIN (5)
Total
Partial

INFORMATION (6)
Up
Down and up
Up, down and up

Numbers refer to the discussion in the text.

Finally, we can identify variations in the *reactions of the participants* (6). What are they expected to react to? Sometimes questions are asked before any relevant information has been broadcast. In this case the background for the opinion is the previous knowledge of the participart. Altnernatively, a reaction on a broadcast programme is explored. Obviously, one may argue whether the programme constitutes a reasonable background, or perhaps a biased one, or even an outright brainwashing. Also there is usually not too much time to digest the information, unless the time to answer is delayed with a resulting decrease in the return ratio. It is possible to organize "upward" information (i.e., incasting), then a programme broadcast "down," and finally to repeat the "up" incast. If certain psychological problems can be overcome, this approach would help us to understand better the impact of our mass media. It is a domain of great opportunities and, at present, little concrete knowledge.

FUNCTIONS

Interactive broadcasting is now performing or, perhaps more precisely, is going to perform quite a spectrum of functions.

One possible function of the upstream capability is the exercise of choice of the message to be broadcast to the viewer. A discussion of one specific business use (pay television) will show the increased importance of this possibility of choice.

Other possible functions, some of which are summarized in Table 5, are usually referred to as "incasting." We shall summarize them here for two reasons, even though they do not bear *directly* on the issue of electronic polling. First, we do have practical experience with them, while there is little experience with electronic polling and voting, and it is possible to draw some analogies. Second, decisions about the introduction of new technologies, and to some extent the public's attitudes to them, will be related to the sum of their existing and potential functions, not only to polling and voting. In other words, it seems likely that their introduction will not be solely for electronic polling; if the method exists in future for this particular use it will be because it is one of many possible uses, the sum total of which is deemed to make the investment in equipment worthwhile. What are some of these multiple uses?

1. *To raise a fast and effective alarm.* The main emergencies covered by this opinion are:

- Fire (to call the fire department).

- Burglary or other security emergency (to call police).

- Medical emergency (to call the physician, hospital or ambulance service).

There are three types of triggering mechanisms:

- Direct manually, that is, via a specific push-button, activated by the customer, who becomes aware of the emergency.

- Remote manually, that is, via push-button on a necklace or bracelet, the alarm being transmitted first to the home console and then the usual way to the centre.

- Automatically. This includes smoke and heat detectors of fire; sensors on doors, windows, and in guarded spaces to identify intrusion; and (for the future) sensors of heart function, temperature, and other body functions to monitor physical well-being of patients or invalids.

The alarm signal may be transmitted

- To a central monitoring unit, equipped with a computer and a hard-copy printer. For example, in case of a medical alarm, the print-out includes the name, address, and main items of the medical history of the patient. For a fire alarm, a description of the premises informs about the power system, location

TABLE 5

POTENTIAL FUNCTIONS OF INCASTING

	Function			
	Control of programme			
	Pay television			
	Alarm: Fire			
	Burglary			
	Medical			
Incasting	Information retrieval			
Incasting	Buying and Paying			
Incasting	Auction			
Incasting	Gambling			
Incasting	Games			
Incasting	Education			
Incasting	Utility meter reading			
Incasting	Audience size estimation			
Incasting	Evaluation of Programs			
Incasting	Commercials			
Incasting	Products and services			
Incasting	Election	Voting	Social incasting	
Incasting	Referendum, plebiscite	Voting	Social incasting	
Incasting	Opinion survey		Social incasting	
Incasting	Advice		Social incasting	
Incasting	Attitude research		Social incasting	

The order in this table is based on their functional similarities.

of combustibles, and so on. Such information is then transmitted to the appropriate final destination, that is, the hospital, ambulance dispatch centre, fire station, or police station.

- Directly to the final destination (police, etc.). Some precautions have to be met in this case. Most police departments, for instance, insist on strict regulations for the quality of sensors and training of operators to avoid false alarms.

- To a local sounding device. This has been found useful to threaten away burglars when there is nobody on the premises and a delay in police action can be expected. On the other hand, it can be dangerous, for example, in a case of a hold-up.

Similar technology can be used for remote reading of utility meters, electrical energy, water, heat.

2. *To receive centrally stored information*. Interactive television, as well as other technologies, may retrieve centrally stored information. In one system, information retrieval basically permits a choice of a page of text that can be transmitted to the viewers' screen after being selected from a data base of the central computer. It may include

- News and entertainment

- Advertisement

- Specifically prepared information (houses to be sold, food guides, etc.)

- General information of an encyclopedic nature

- Educational material (which in turn may use incasting during the use of the text).

3. *To conduct an electronic transfer of funds between banks and major customers*. In both Canada and the United States, this is becoming a growing mode of payment. To a very limited degree, interactive broadcasting can participate in this mode when viewers actually agree to make a payment that has just been suggested on the screen--to buy goods or services, bid in an auction, gamble in a lottery, or participate in off-track betting. From this technique, quite a number of important issues arise. Some of them have been met before with credit cards: limits on the liability of the customer, including cases of negligence; selection of viewers who can buy; psychological pressure from easy credit and postponed payment. But other issues are related to the technology; identification of the person (who can use the terminal for making payments?); determination of whether it is a legally valid act (confirmation of approval, since pushing a button is not perceived as equivalent to a signature on a cheque); confidentiality; and detection of errors and frauds.

4. *For amusement*. Video games are a growing and heavily competitive business. Traditional television has a tendency to isolate an individual from his

environment and keep him glued to the screen. Television games, on the contrary, are bringing together at least some family members or friends. Interactive television invites viewers to play against the centre, for example, to answer quizzes and test questions. As far as we perceive it as human interaction, it is an improvement in the "lonely crowd" phenomenon. If we feel that it is another dehumanizing gimmick, replacing human interaction by a fake, it is the contrary. Potentially, there is another possible gaming situation: *ad hoc* groups, of their own choice, playing together without regard to their place of living. This option depends on the commercial use of technologies for general information systems and will therefore have to wait for a short time, although a simplified version has already been tested.

There are other uses of the system which the operators would find attractive but which would be of little interest to viewers.

5. *To obtain ratings*. Television programming as we know it turns around ratings. How many people have watched a particular programme? How much is it worth to business for commercials? How much does it increase the competitiveness of a cable system? A.C. Nielsen & Co. has pioneered a traditional and reliable survey process for determining these answers. As it is rather slow, devices are now increasingly being installed in television sets to register responses of a sample of viewers. Interactive television can estimate the number of viewers within a few seconds by monitoring which channel is switched on. As not all viewers are plugged into the system, problems of sampling and representativeness still have to be solved. Nevertheless, direct electronic rating seems to be, for the near future, a major application of the technique.

6. *To evaluate programmes*. Audience estimation is one way of judging how viewers like a programme. Interactive television makes more direct judgement possible, either by asking occasional questions, or giving the viewer a permanent opportunity to express his opinion. Still more important is the possibility to judge opinions about commercials. Marketing professionals have always had problems with feedback from their advertisements. To register the immediate reaction of a viewing sample just after (or even during) the commercial is a marketing expert's dream. In addition, consumer panels have been used for a long time to comment on products that are on the market or being prepared for introduction.

If interactive television were available, at least some of the features of a product might be evaluated by a sample of viewers at less cost and with less delay.

To touch upon the potentiality of interactive television for evaluating programmes is to approach the issue that is the main focus of this essay: the use of electronic media for political and social polling. The lower part of Table 5 lists functions which more closely relate to the concerns of the present text. Here they are referred to as "social incasting" functions. This general term applies to channels that transmit to the centre information about social and political matters. The range of functions is considerable. One alternative is the collection of informative communications, those which more or less influence a formal decision to be taken, say, by Parliament or Cabinet. In these cases the viewer might be asked "What, in your opinion, is the main cause of inflation?" In another case, explicit advice is sought. For example, viewers are asked whether a particular zoning regulation should be withdrawn. It would also be possible to use the technique for a more complex analysis of a cluster of attitudes towards a certain problem or policy, thus going considerably further than merely identifying a simple opinion. However, this would require concurrent application of other techniques and there are no known indications of any concrete steps in this direction. A further potential use of the technique is for recording a vote: to make a choice of representatives, as in a election, or to express a view on any issue, as in a referendum or plebiscite.

There are two other similar functions that fall outside the domain of this study. One of them is the use of specific electronic channels for public voting in general elections and referenda, where the traditional paper ballot would be replaced by switching a circuit, but the same publicly supervised process would be used as now. The second is the counting of votes in a meeting or assembly (e.g., in Parliament) by electronic means. Both these functions use technologies specific to the circumstances and are not related to the broader social and political questions under discussion in this essay.

CHAPTER TWO

The last chapter has summarized the main functions that incasting can perform and the main technological principles on which it relies. Let us now review selected examples of concrete applications.

There is a lot of hardware around, from Telecable Videotron in Montreal to Rapidflex in Japan; thousands of phone-ins and phone surveys; still more ideas about what could and should be done; and an infinite number of systems and developments that may have marginal effects on social incasting. The following paragraphs are a selection from the known ones. It is hard to say how many have escaped the research net.

QUBE

Qube is a thirty-channel interactive cable television system operating in Columbus, Ohio. Its upstream capability uses as input a small console (see Figure 2). A button at the bottom chooses the column and one of ten buttons on the left side chooses one channel. There are three columns: "Premium" or "p" channels, with a yellow button, offer a choice of sports programmes, first-run movies, classics from the past, present drive-in programmes, concerts, and educational courses. There is a fee for the programmes: classics come at about $1.00, first-run movies for $2.50, soft porno for $3.50. There is a two-minute grace period in each programme when the customer is not billed, apparently to permit sampling and channel selection. Certain programmes can be viewed for one fee up to four times during the same day.

The "T" channel, with a blue button, offers the choice of ten television stations and is free. The second column on the console, "C," with a green button, is for community programming. These programmes are also free. They include consumer-oriented issues and comparisons of prices in grocery and department stores("Mr. Qubesumer"); programmes and discussions about the activity of the city hall, for example, about city by-laws under consideration, or about the work of the mayor.

While watching the programmes viewers can use any of the five black buttons at the right hand side of the console to answer questions that have been posed to them. Do you like the programme? Should the programme continue? Do you agree with the proposal? Alternative answers and numbers of respective buttons appear on the screen above a flashing instruction to "touch now."

One of the topics in the community programming is consumerism. "Mr. Qubesumer" has shown that the "miraculous no-hunger bread" is a fraud; that "vine-ripened" tomatoes have been "gassed pink" in the nearby warehouse; that the bowl of soup at a local restaurant costs twice as much as a cup, but holds the same volume. He also compares prices regularly in local grocery and department stores. Representatives of a government agency and a state attorney have even

FIGURE 2

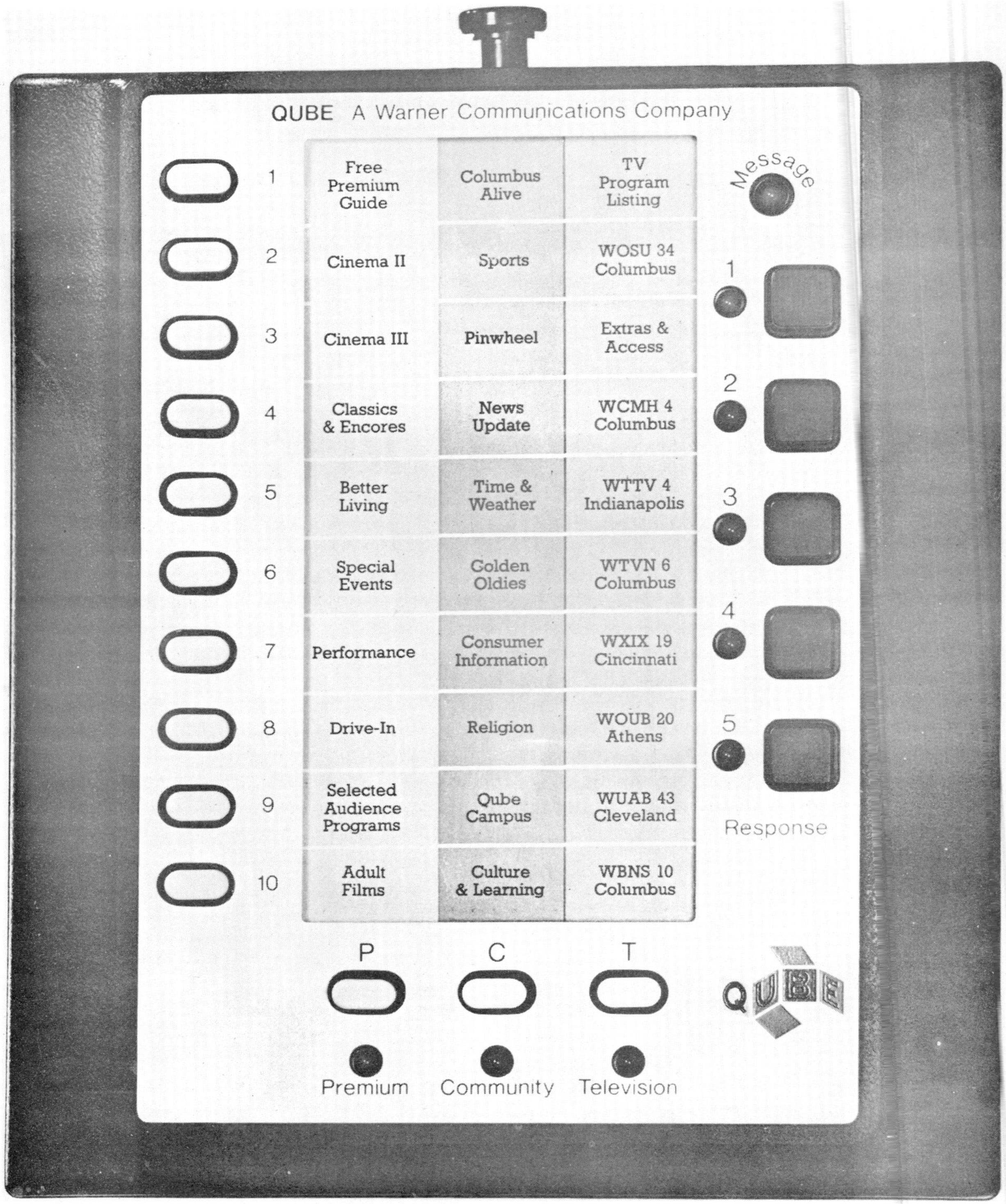

User console in the Qube system (150 x 175 x 60 mm). Five push-buttons for incasting are on the right.

been invited to a show in which viewers have been asked who feels cheated by a company, who sells flour for the "miraculous bread" and who would take part in investigations, (the culprit company is now out of business).

Another offering of the Qube service is "Talent Search," in which local residents get 150 seconds to show whatever they wish to show: singing, dancing, music. If there are more "no" buttons pushed than "yes" buttons for 40 seconds, the performance is stopped and another talent gets his 150-second chance.

Daily from 6:00 to 8:00 p.m. is the time for the popular "Columbus Alive" programme. The co-host is Susan Goldwater, a fashionable blond who recently separated from representative Barry Goldwater. In a discussion with John Dean, 53 per cent of her viewers answered that they would accept a dinner invitation from Richard Nixon. Only 49 per cent, however, liked Susan Goldwater's new hairstyle, and nearly all rejected a toupee modelled by her new male co-host. In other shows, viewers have voted on issues ranging from the way the city mayor handled a snow emergency to the major causes of inflation--(1) government spending? (2) energy problems? (3) Russian wheat sales? (4) wage demands? (5) others? Touch now!

There seems to be an unlimited supply of other ideas: education; games (volunteers, selected at random, compete by telephone); general research; a possibility to buy merchandise or services, or to order sales information; opinion polling about a new magazine cover; even a sixteen-hour-a-day preschoolers programme. In addition, fire, burglary, or medical alarms can be triggered by the console or by a miniature sender from a distance of fifty miles.

The heart of the system is a central computer "Eclipse," able to handle 100,000 subscribers. It checks every subscriber once every six seconds to find out if the set is on, which channel is requested, and if any of the five "answer buttons" has been pressed. According to this information, it computes fees for programmes and bills every customer once a month by mail. The second function is to summarize and tabulate the ratings, that is, to count how many customers view any channel at any given time. Its third task is to register and tabulate the touching of answer buttons. In the normal operation mode the identities of answering customers are forgotten as soon as summaries are tabulated. Occasionally, for example, if pushing a button indicates a sale or request for information, identifications are printed out.

Qube is the brainchild of Warner Cable Corporation, which owns 138 other cable systems in the United States. It is wholly owned by Warner Communications Inc., one of the "big ones" in motion pictures, television production, recorded music, publishing, toys, sports, and video games. The consoles are built by Pioneer Electronic Corporation in Tokyo and have been tested since October 1975. Regular programming started 1 December 1977, in Columbus, Ohio, on an average cable network with 26,000 subscribers. Within several months, about 20,000 of them chose to join Qube for an installation charge of $9.95 plus $10.95 monthly fee. The number of subscribers is increasing but company officials will not say more than that they are "very, very encouraged." The original cost estimate of $8-10 million has jumped to $15 million, according to the company, or to above $25 million, according to outsiders. Extension to other locations is now under way.

The Qube experiment is regarded as a success by most viewers, users of information and competitors and, of course, is described as a success by the company.

Technically and, from most points of view, legally Qube could be introduced into Canada tomorrow.

VOTAPHONE

Bell Northern Research has developed in its Ottawa laboratories an interesting incasting system using the standard telephone network for upstream direction. A simple console is located near the residential television set and connected to the local phone. Two push-buttons on the console are labelled Plus and Minus. In the most simple case, they may indicate answers Yes or No respectively if pushed during a 10-second answer period. There is a possibility to indicate up to eight answers, if the period is divided to four 2.5-second intervals; pushing buttons in the first interval indicates answers 1 or 2 respectively, in the second interval answers 3 or 4, and so on. There are safeguards against fraudulent, repeated voting.

The console is connected to the telephone switching centre by longitudinal connection, making the use of Votaphone functionally independent from the standard use of the telephone (see Figure 1). The processing of incast signals in the centre depends on the type of Votaphone used. There are two of them: the S-type Votaphone is distributed to a statistically controlled random sample of population in the geographic area and the main characteristics of each household are registered in a central computer. Each vote from an S-Votaphone, is identified in the local telephone switching centre by a magnetic core, brought to the central computer, processed, after which the results are printed out. Unless there is an agreement with the respondents in a specific case, the identity of respondents is not revealed.

The U-type consoles are available to any interested customer. About one hundred potential customers may be served by the same integrating device (magnetic core) in the local switching centre and a cumulative signal is again recorded by the central computer. As the identity gets lost in the switching centre, anonymity is practically guaranteed.

When a number of impulses arrive simultaneously, some may get lost due to the physical limits of the circuit. This can cause an error. Analysis has shown that the accuracy of recording has a statistical impact somewhat similar to the sampling error in survey work. Even in the worst case the error, within acceptable confidence limits, is small enough to guarantee sufficient accuracy. From the engineer's point of view, Votaphone is a neat and clever work. It also looks as if it could be inexpensive.

Votaphone has been designed as a voting and polling mechanism. It can be (but has not yet been) adapted for audience size estimation, automatic meter reading, pay television billing, and alarm reporting. As it uses the phone, it is not dependent on the television cable. The consequences of this difference could be considerable, if, indeed, the system is ever introduced.

As early as January 1975, a paper on the Votaphone was read at the annual meeting of the Society for General Systems Research in New York City. It triggered widespread discussion, along with the expectation that a feasibility trial to test the concept under practical conditions would follow. And yet, to date, there has been none, despite the fact that the system seemed both technologically and economically feasible. Why?

One possible line of speculation is based on the organizational structure of Bell Canada. The research and development belongs organizationally to Headquarters, while applications (including a test application) are a regional responsibility, most probably of the Ontario region. Such an organizational dichotomy has occasionally caused delays in many, or perhaps in most, companies, using high technology. On the other hand, the company's track record in this regard is not too bad.

A more probable reason relates to the fact that Votaphone has been perceived from the beginning as a "voting and polling" device, while comparable developments in the United States have been mostly oriented to pay television ratings, and entertainment generally. In 1975, and perhaps even now, the concept of "citizen participation" suffered a poor reputation in many business circles; it might have been extended to a technology for popular polling and voting. A good part of the bad name was probably based on semantic misunderstandings and naive proclamations made in the late sixties and early seventies. Whatever the reason, it is hard to blame a company, even in speculation only, if it avoided what it perceived as a "hot" issue. As it stands now, we may not even be able to find out if Canada has lost an opportunity, and to learn from it. It is possible, of course, that its introduction was affected by the emergence of the videotex system.

VIDEOTEX is the generic name for interactive alphanumeric and graphic communication equipment developed for the broad population. In February 1979, Bell Canada introduced its pilot videotex project that may soon make publicly accessible a vast amount of information on a standard television set. The system is called Vista and consists of three components: a simple terminal with a hand-held keypad for each user who wants to read selected items from a central data base; a console for each company or agency wanting to supply information into the central memory; and a computerized database control centre with the central memory. Developed by Bell Northern Research and using the information content created by Southam Inc., Torstar Corporation, and Tele-Direct (Bell's subsidiary), Vista employs a standard telephone network (Figure 3).

FIGURE 3

PILOT CONFIGURATION OF THE VISTA SYSTEM

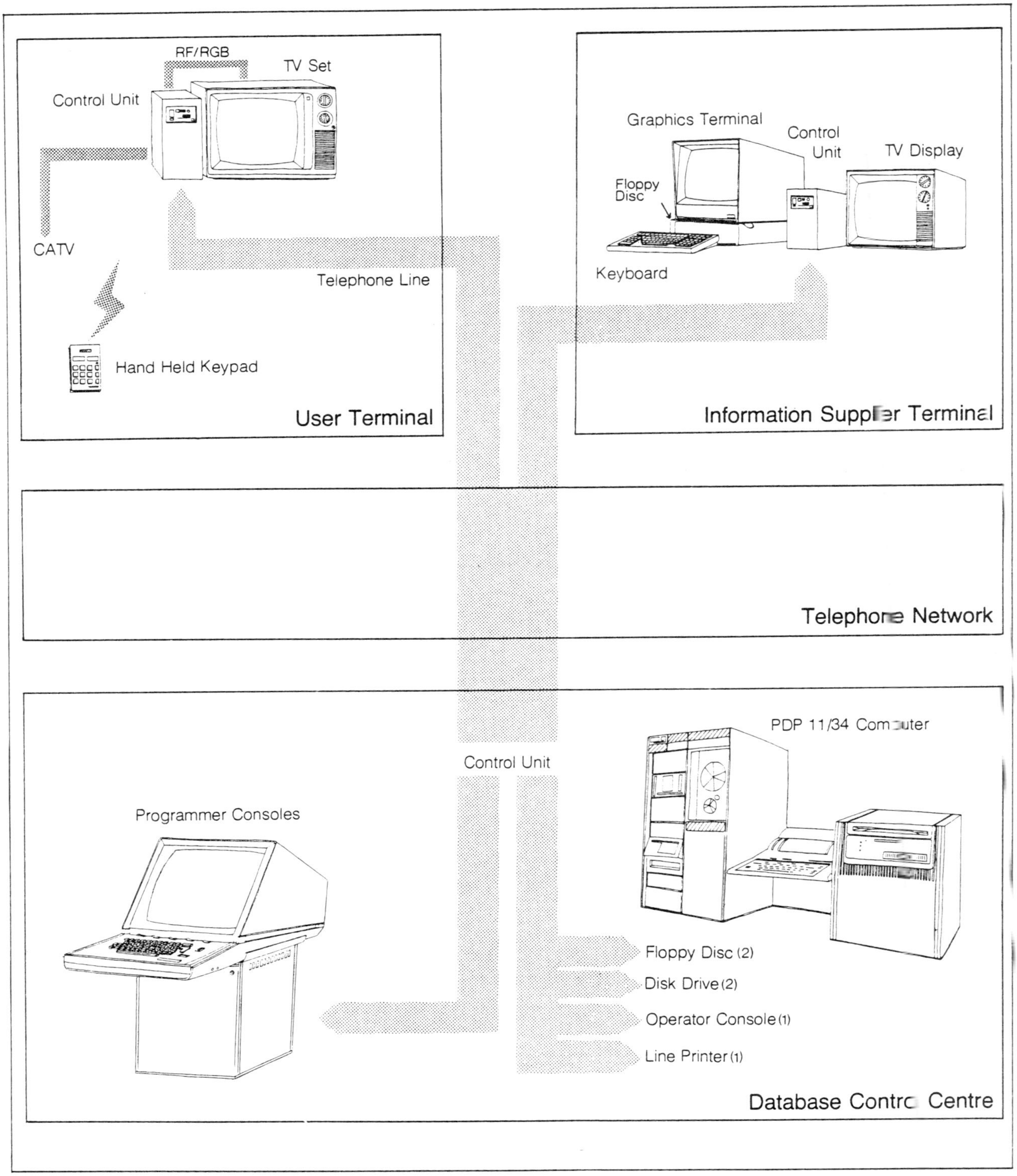

The main features of the system are as follows:

- Content is organized in a tree/menu structure.
- Pages may be accessed either through tree search or directly by page number.
- Content may be created, stored, and displayed in both French and English.
- System-descriptive information and operating instructions are provided to aid the user.
- Content will be added, changed, and deleted as new applications and demonstration packages are developed.
- Content is represented and accessible by three directories which provide a table of contents, subject index, and information-provider index.

The pilot system contains the following: news, weather, sports (national news, international news, sports news, weather, financial news); consumer information (grocery shopping, mortgage calculations); entertainment (movies, restaurants); fun and games; classified ads (new and used cars, real estate); and yellow pages (hotels, contractors' equipment, employment agencies).

At the beginning of the experiment, twenty-five users' terminals made accessible 2,000 pages of text in seven colours, in English or French. The pilot project will run until spring 1980, to be followed by a large-scale concept test and commercial service in 1981 or 1982. The cost for the future user is not known but will consist of component charges (1) for Vista user terminals, (2) for access to the Bell Canada Vista service, and (3) for access to specific information items.

During 1979, a new improvement in Canadian videotex was being tested in conjunction with Vista. Telidon, developed by the federal Department of Communications, is a sophisticated information system technologically independent from the terminals. At any given time the terminals which are used (at present television sets) will put some limits on resolution--the present picture being relatively rough--but the Telidon approach makes it possible to increase the resolution when the quality of terminals and transfer channels improves without fears for dynamic compatibility (i.e., the possibility exists to use both the older and the new equipment). In this respect, Telidon is superior to other videotex systems.

The system is to be tested in 1979 in Ontario and Alberta (120 homes in Calgary tested by the Alberta Government Telephones). As compatibility between Vista and Telidon is uncertain, a choice will probably have to be made one day. Some day one of the two may well complement or replace specific, simple incasting systems in Canada, if they should actually prove to be ready earlier.

Basically, however, videotex is a much more complex and considerably different system than the above-mentioned incasting systems.

The estimated price of $200-$300 for the terminal is perhaps optimistic, but not unrealistic. Prototypes now cost $15,000, but Norpak Ltd., operating out of Pakenham, Ontario, is trying to get the price down for mass production. Economically, the development represents a potentially big market for Canada, although not as great as some semi-official statements or subjective guesses of persons who have reasons to wear rosy glasses would suggest. The idea that "five million Canadians will subscribe to videotex in five years" is patently impossible. A more probable estimate would be 350,000 videotex terminals sold in Canada in 1983, if estimates by British manufacturers for the United Kingdom are taken as a basis. All in all, it could (in the best case) mean between $0.5 and 1 billion business within the next ten years depending on market penetration. (For comparison: $3 billion worth of electronic equipment was purchased in Canada in 1977, two thirds of it imported.)

For its normal operations, videotex faces one potential obstacle: the switchboard centres of a telephone network have been developed for the "dialogue pattern." If videotex achieves deep penetration, potential overload from the "broadcasting pattern" will cause the extension of some switching centres to handle a heavy flow of traffic. This, in turn, will cause more investment in the centres (unless, in the extreme case, it were to bring competition from other technological systems).[2] On top of it, incasting would cause an additional overload at peak time, that is, immediately after a question has been asked. Success of videotex in incasting will depend on, among other factors, economical solutions to both aspects of the overload. The expressed hopes for success are high, but no satisfactory analysis of the problem has yet been found in the public domain.[3]

TELEVOTE

During the Hawaii State Constitutional Convention in August 1978, a relatively new survey process was tested by the University of Hawaii in Honolulu. A random sample of 560 citizens was selected, 500 of whom agreed to participate, 400 of whom actually did so, and about 350 of whom volunteered to continue their participation. Each respondent received for each vote a four-page explanation of the topic of the vote, for example, "How to Select the Best Judges for Hawaii." The explanation summarized possible alternatives and tried to include pros and cons for each of them. At the end, several closed-end questions were included, of which the following is typical:

> Now, on the basis of what you have just read (and what else you might know or believe) which method do you think is best for Hawaii?
>
> -Appointment (Hawaii's present system)
>
> -Election of judges (who must run on their own and not as a member of a political party).

Participants reported their answers within three days by telephone to the centre which immediately published the popular results, complete with frequencies, cross-tabulations with demographic variables, and comments on the sampling and its impact.

Televote and its centre are run by university students as part of their curriculum under the guidance of Professor Ted Becker who pioneered the idea in Hawaii. Before being tried in Hawaii, Televote was used by Vincent Campbell in the San Jose School District in California, in 1973-1974. In Hawaii, Televote will almost certainly continue, as the public reaction has been generally favourable. Preparations for other applications are under way.

The limits of the approach are rather obvious: the information transmitted to the subject is limited and no one knows how much the participants read and understand. Both the number of questions and number of choices are limited. In addition, there are always some doubts about anonymity when volunteers are involved. Obviously the scale of the survey is limited by the number of students and cannot be too much further extended as long as it is a part of the students' curricula.

On the other hand, Televote seems to be a success. People say that they trust a university more than they would a government agency or an institute. Moreover, the results have been welcomed later by politicians and the mass media, and the people, in general, seem to like it. Further, it appears to provoke interest and discussion beyond the sample population, and at a very reasonable cost.

TALKING BACK

CBC Television in co-operation with the Canadian Association for Adult Education broadcast, during the winter of 1978 and spring of 1979, a series of programmes described as "cross country conversation involving millions of Canadians linked together by television, telephones, and computers, and by a network of local, provincial, regional, and national organizations." It included one three-hour (prime-time) programme and five half-hour programmes in two- or three-week intervals. The topics were

- An Experiment in Democratic Sharing
- So We're Different!
- Let's Get It Straight!
- Winners or Losers
- Canadians at Work
- Where Do We Go from Here?

The programmes were broadcast from studios around Canada with direct connection between them, making some immediate discussion possible. Phoning in to provincial centres during and after each programme was encouraged and some calls were selected for broadcast: discussion groups of two to twelve people were invited to register and to send reports after each programme by telephoning collect to provincial centres. The main part of each report consisted of three recommendations for action, each not longer than five words. Recommendations and other comments were summarized and sent back to groups before the next programme.

The first programme triggered about 1,500 calls and the creation of 233 groups. In all, some 1,728 participants reported on a broad array of problems. As with any such approach, there were both positive and negative comments. People in general and the media in particular criticized the fact that most opinions in the programme had been heard many times; that the percentage of participating people was low; and that the statements of participants were naïve, non-representative, or predictable. Other people and media praised the programme as innovative, offering a chance for everybody to send his/her message; as a step towards democratization of the mass media.

HI-OVIS

An interactive television system using optical fibre was introduced in July 1978, in western Japan. A group of 158 homes was chosen as the first "model township" or "monitors" for testing the system. Each monitor's home is equipped with a special 20-inch colour television receiver, terminal adapter, camera, microphone and light receiver-transmitter. By pushing keyboard buttons, the operator can choose from any of twenty-nine channels constantly available, including urgent medical or local event reports. He can also ask questions and get instant replies, or supply information through a camera and microphone. In addition, citizens who have not been designated as "monitors" are still able to participate through equipment placed in fifteen public places.

The purpose of the installation is testing. After the first experiments in Tokyo, the district of Higashi-Ikoma in the Napa Prefecture was choosen (the name Hi-Ovis stands for "Higashi-Ikoma Optical Visual Information Systems"). Several up-and-coming companies are working together: Sumitomo Electric in optical fibre transmission; Fujitsu in computers; and Matsushita in audio-visual technology. A broad range of small and large components is being tested: community studios and mobile studios for local programming and citizens' input; character generators for two types of Japanese scripts (hiragana and katakana); standard alphanumeric characters and simple diagrams, used for rapidly changing information (e.g., weather forecasts or business information); microfiche-and-audio equipment for semi-permanent memory such as for dictionaries and education programmes; computer-controlled video-cassette equipment; and, finally, the channels with numerical keys and microphones, used by subscribers to answer questions that have been asked from the broadcasting centre.

At present, the tests seem to be proceeding well. If they are successful, a nation-wide system could be in operation in 1985--or so the Japanese hope.

OTHER EXAMPLES

In 1975, a two-way television system was introduced in The Woodlands, a new community north of Houston, Texas. Its backbone has been a set of alarm channels--medical, police, and fire--developed by TOCOM (TOtal COMmunication) of Dallas. One thousand and one hundred Woodland residents enjoy a 25 per cent reduction in insurance rates because of the system. Concurrently they obtain cable television and can use the system to vote on community issues and hear about several other forthcoming improvements. The cost of the user's console, $300, and a share of the $150,000 cost of the centre were included in the price of each new house. At least eight other communities, including Dayton, Ohio, are using or are ready to use the TOCOM system. Broward County Cable Co. is starting to spend $9.2 million to lay 600 miles of lines in Fort Lauderdale and Oakland Park to introduce the system. Broward County Cable Co. and one third of TOCOM are owned by Selkirk Holdings Limited, a Canadian company.

Roger R. Percy in Seattle, Washington has introduced an interactive system, that has been developed for him by SRI International in Menlo Park, California. A miniature user console has six buttons, described as "zap," "funny," "I buy what you're saying," "you're conning me," "more information," and "applause." Viewers push buttons whenever they like, without waiting for questions. Their answers are transmitted by a loop over telephone lines, together with the information as to which channel is switched on to the centre. Viewers seem to like the idea. So do advertisers who learn fast about the reactions of the public. In so far as this system is comparable, the results of this spontaneous response system have been very similar to traditional surveys done following their programmes. However, the method of incasting brought in more information faster.

During the early 1970s, experiments were started in the new town of Tama, north of Tokyo. There, selected households can now use standard touch-tone telephones to request programmes and retrieve data on their television sets. The Tama experiment has served as a stepping stone to the Hi-Ovis system, described above.

The Ministry of Posts and Telecommunications, which pushed the work in Tama, has been interested in the communication aspects, while the Ministry for International Trade and Industry (which is behind Hi-Ovis' experiment) is more concerned with the manufacturing potential of the new hardware. However, both these experiments have to be regarded as components of an intensive effort to develop an improved communication network in Japan, including

- Facsimile transmission. This has been delayed for some time by government regulations banning this type of transmission from Hertzian channels, but is now growing very quickly in the business sector.

- A data retrieval system. Delayed by strict limits on cable television, *Nihon Keizai Shimbun*, Japan's largest business daily newspaper, has now launched a prototype using its own data bank.

- A nation-wide high-speed, high-capacity satellite-based network. The Ministry of Posts and Telecommunications would like to see the satellite in existence by the early 1980s, but there are problems with the full use of its capacity: nobody wants to reconstruct the television network and facsimile transmission will not yet generate enough traffic; moreover, there is no rush to introduce other mass-volume applications.

In relation to incasting the development in Japan has three dimensions:

- The regulatory obstacles, as mentioned above, have been greatly reduced.

- The technological development of mass communication is, according to most observers, ahead of all rival countries.

- Little is known about the human and political role. Nippon Hoso Kyokai (the Japan Broadcasting Corporation) is doing a creditable job in public opinion research, but has not yet released any results concerning incasting. There are people who believe that some specifics of the Japanese culture could very well co-exist with the incasting technology.

Taking all this into account, it would not be surprising if, in the near future, we have to look to Japan in order to learn more about incasting.

OTHER DEVELOPMENTS

There are several technological improvements, ready or in progress, that could have impact on social incasting.

Fibres have many advantages over copper wires as signal carriers. There is, however, a very prosaic one which can be important: they require less space. Excavated space under our cities is in most cases overcrowded, and more digging would be bad for most concerned. A Canadian experiment started in December 1978, when thirty-five houses in Yorkville, Toronto, got their fibres but the introduction will be slow. The technology could, however, make one hundred channels available to every customer and scatter the market. For incasting, it would mean more potential opportunities to participate, more competition and fewer participants per incasting channel.

There might be competition for fibres, however. About three thousand amateurs in the United States and Canada have already spent $6,000-15,000 each to build dish antennae and receive signals directly from satellites. The Japanese may soon offer a dish antenna for $200. On the other hand, routine use of the direct link is improbable for business reasons. If its use were to spread, it would bring quite a lot of new problems to incasting, some of them difficult to solve. (There is, however, a new threat to communication satellites. Some Third World nations near the equator claim they own the satellites parked overhead and the United Nations seems to agree. The claimants are considering buying a rocket missile that is being manufactured by a German company in Africa and which is technically able to destroy the satellites.)

Further, voice communication may one day replace push-buttons. A terminal that can recognize sixty words was sold in 1978 by Threshold Technology, Inc. for $10,000. The price is expected to drop to $5,000 during 1979 and to $500 by 1985. "Large vocabulary natural language systems" are making rapid progress in laboratories and may be on the market by 1985. For incasting purposes, however, their price will be prohibitive for a long time.

It would not be reasonable to regard such technological options as improbable or irrelevant dreams. In April 1979, Tagramiut Nipingat (the Inuit Communication Society of Arctic Quebec) received a $1 million federal grant to prepare the technology for both standard television and incasting. For the first time, town hall-type meetings in scattered Arctic communities will be possible, using communication satellite Anik B.

SPECULATION

What can be the impact of social incasting? This question is hard to answer in a reliable way as our experience is too limited to serve as a basis for a confident forecast.[4] There exist, nevertheless, the first experiences with some of the other functions of interactive television--choice of programme, evaluation of programmes, and so on--and the following paragraphs therefore use these to speculate about the social impact of electronic polling. The speculation is based largely on one case (the Qube experiments in Columbus, Ohio, during 1978) and on one source (newspaper and journal articles). The limited sources permit only a series of impressions and, obviously, restrict the value of the conclusions as firm predictions. But the issue is worth the risk. Better to be off the bull's-eye than not to see the target at all!

* * * * *

"Pinwheel" is a children's programme shown on channel C-3 from 7:00 to 9:00 P.M. Dr. Vivian Horner, the vice-president for Educational and Children's Programming in the company, is a distinguished and experienced educator, past director of research for the "Electric Company" children's show. She is very sceptical of programmes like "Sesame Street" and perceives them as pushy, too oriented to measurable goals, too much inclined to use advertising techniques. "Pinwheel" is different, she says: it is soft, gentle, awareness oriented.

This is an ideological question. Dr. Horner may be right or wrong. Either way, her ideology (i.e., the vice-president's ideology) will prevail. We cannot conceive of any push-buttons or communication patterns that would involve the public (not even the adult public) in the control of the general direction or ideology of the channel. On the contrary. As far as the political orientation is concerned, the public may be so busy pushing buttons on questions of low relevance that they will forget to ask what they think about the channel's orientation as a whole. In turn, the general orientation will influence the public's perception and taste. The message from the public can be interpreted in two ways: either "how to give them what they want" or "how to give them what we want in a palatable way." Nobody knows what will be the position of future broadcasters between these extremes.

* * * * *

Channel P-10, known affectionately as "Porno," brings adult movies, carefully edited to a soft-core level. (Quite a number of viewers complained that important scenes were omitted from the "Swinging Stewardesses." The general level of relevant information seems to be higher than expected!)

As much as the viewers may enjoy the programme, however, they tend to dislike an explicit record of their choice on the monthly bill. It is one thing to invite good friends for a favourite "P-10 party"; it is another thing to face the risk that somebody will abuse and publish the records (who believes in confidentiality of records anyway?).

What if it were to happen to a candidate before an election? "If you are afraid of being blackmailed, do not watch blue movies," says G.M. Hauser, the Warner Cable chairman. This may be good advice concerning porno. But what about information about the candidate's viewing patterns generally? About another person's viewing pattern? How do we know (as distinguished from what we are being told) who might record our choice of programmes and how it will be used?

The console in the Qube system has a "pay-key"; it can be removed to guarantee that no one will view pay movies on our television set. Shall we one day remove other keys from our console to guarantee the proof that we are not watching the wrong programmes--perhaps radical-key, or redneck-key or opposition-key?

* * * * *

During the "Talent Search" programme, local talents get 150 seconds each to show their ability--unless 50 per cent or more of the audience vote "no," in which case they go off the air immediately. The reasons for dismissal are irrelevant. Even if an accordion player provokes criticism because of his "accent" (he happens to be an Italian), the message on the station is "they do not like him; let us give them something else."

This approach works well for the selection of programmes people like to watch. But what about our attitude to future political decisions? What are we going to vote on? An offered solution? The wording of the programmes? The broadcaster's voice? If a kid from Jones Junior High is on the air, all the kids from Jones vote, says a young viewer; the same is true for their main rival, Hastings Junior High. Will pressure groups organize "gang voting" in a similar way? Such a development could bring into play the organizational abilities of a group, certainly a politically relevant factor. The outcome, however, would be a fuzzy result of organizational abilities and opinions. How should it be interpreted? How will it be interpreted?

* * * * *

And what about the timing of the programmes? Not too many viewers paid $2.50 for *I Pagliacci*, when a comparable opera was transmitted from La Scala free at the same time. What about airing, by purpose or accident, an opinion survey concurrently with a major hockey game? Or at the same time as an important political programme? What will be the impact on the selection of participants? One can imagine less obvious and less identifiable impacts of clever timing.

* * * * *

Qube viewers have indicated in surveys that they want more television sports. However, when they actually got the choice they did not use it much. Perhaps it is different to answer a question in a survey (which is free) and to hit a pay button (which costs $2.00). The Ohio State University basketball team first won four times, and fans watched and paid. After two consecutive losses the audience nearly disappeared. We find it difficult, it seems, to predict who will watch a basketball match. How well are we going to predict, through the same mechanism, the political behaviour of citizens?

* * * * *

Who really does decide to "plug in" anyway?

An authoritarian mother instructs her children to push "No" in answer to the question "Do you favour the building of a sports arena?" There has been no discussion and it is not clear what is the opinion of the children who, after all, are potential users of the arena. A little later some questions become embarrassing: "How many of you have smoked marijuana?" (83 per cent). "How many of you know anyone who takes Valium?" Perhaps to prevent discussion on topics covered by a family taboo, the mother finds her presence is necessary elsewhere. The children are left to answer for themselves. Within seconds the microprocess of decision making on "how to vote" has changed.

Who, in this actual case, really voted? How was it reflected in the results? A Bell Northern expert, presented with this case, suggested that the strongest character in the family decides--both in interactive television and in life. The results may therefore be more realistic than we think. Perhaps he is right. Perhaps he is not. We do not know how the decision-making microprocess works. We do not even know how to find out. How shall we interpret the results in socially relevant cases?

* * * * *

Against all expectations, television seems to be going through a path similar to that taken by radio earlier; from centre of interest to background. An increasing percentage of viewers keep the television set on while absent from home or talking, reading, or playing. Reactions in incasting can be interpreted only if we know what it is a reaction to, what the viewer has just seen and heard. How shall we be able to find out, when the information that "the set is on" is obviously insufficent?

* * * * *

Columbus, Ohio is a test city. The composition of its inhabitants is assumed to be typical to that of the United States. What goes in Columbus, say the marekting experts, goes in the United States, (London and Sherbrooke are sometimes used in a similar way in Canada). On the other hand, frequent testing has been found, on similar occasions, to have considerable impact on attitudes. A "professional panel member" and a "professional panel community" are two of many medium-size nightmares of marketing. Question: how far can we make conclusions from the incasting-related behaviour of Columbus viewers? How far does it hold for the entire United States? How far do any such U.S. conclusions hold for Canada?

Qube programming vice-president Harold Kleiman is careful to avoid the saturation point, beyond which the viewer would say, "We'll pay you anything but please, please do not ask our opinion any more." Mr. Kleimen is no political scientist, but he knows his audience pretty well. Could he be right even about the approach of citizens to any input (not only via interactive television) to problems of governance? To public problems? To a so-called "ideal government" which takes broad consultations seriously? How much do people really want to be heard?

* * * * *

During a football game, the viewers had an opportunity to give their advice to the coach. There was little enthusiasm for this version of incasting after the game. Reason: the coach *did not use* the viewers' advice; they felt that their opinion had been treated as irrelevant. How will it be when the viewers use the upstream capability to give advice and opinion to the government? Will anybody consider it? Will the viewers believe that anybody does? How will they react if they do not believe that it is considered?

* * * * *

These last speculative paragraphs do not, obviously, provide an exhaustive summary of the problem nor final answers as to what the impacts of social incasting might be. When we do have final answers, it will probably be too late to do anything with them. Some of the questions will seem, with the benefit of hindsight, paranoid, and some answers plainly wrong. We will have to keep our minds open, as our guesses are not always too reliable. Who would have thought, ten years ago, that the female viewers of a television programme in a U.S. community would be asked through the television set in their homes to answer the question of whether or not they regard their mate's penis as too small? (Twelve per cent answered "yes.")

CHAPTER THREE

As suggested earlier, it seems unlikely that electronic incasting will be widely available just because it would permit social and political polling. Other uses will be required before people will justify the purchase of the electronic package. But once available, it seems reasonable to forecast that pressure will develop to use the technique for social incasting. This is because the present means of obtaining information on the views of the general public are widely regarded as inadequate.

Pressure is frequently felt in many ways. The increasing frequency, for example, of calls for direct democracy can be taken as a sympton of an increasing distrust for, and frustration with, our political institutions. The following paragraphs summarize briefly several dimensions of the pressure that are frequently discussed but seldom, if ever, documented.

Members of Parliament find it increasingly difficult to keep in contact with their constituents. At the federal level, long parliamentary sessions limit the M.P.s' time for personal contacts in their ridings, and the use of assistants in the constituencies or secretaries in their Ottawa offices, while generally helpful, contributes further to the isolation.

The problem of remoteness from the attitudes of ordinary citizens is even more acute inside the public service. For many reasons, one of them being the complexity of current public policy, the size of the public service has grown considerably, a fact which does not help create good communications. More significantly, the decision-making pattern during the policy elaboration process has changed. In the past, it was mainly a two-step process with decision making in both steps in the hands of the deputy minister. First, he decided what information he needed and requested it. Second, he decided on the policy and requested its elaboration, always keeping in mind what other tests the policy would undergo in the future. This pattern is occasionally applied even today. However, most important policies are arrived at by a different route: the decision, "what policy should be developed," is partially pre-empted by the preceding decision, "what policy should be elaborated on." This decision is in turn partially pre-empted by still another decision on "what components and aspects of this policy should be elaborated on."

As far as the information about the public's needs and general attitudes is concerned, there were difficulties even with the simpler, one-decision-level situation. In the present pattern, the vertical flow of this information has virtually disappeared. There exists a heavy horizontal flow of information and advice between different departments, but it is seldom much concerned with the attitudes of the public. The result is the increasing incapacity of governments to foresee popular reaction to their decisions. Repeatedly, there is surprise when popular reactions to government actions are unexpected, inexplicable and contrary to the stances taken by the major actors in the policy-making process. This, in turn, has the paradoxical effect of driving the public servants to perceive the feedback they get from the general public as less and less important as compared to other, narrower criteria and other, more immediate reactions.

Research, integrated into the policy process, may be an effective contribution to the process of understanding citizens. Unfortunately many of the existing research projects use inappropriate (frequently obsolete or pseudosophisticated) techniques or are isolated from the policy process. When a politician who dreams up a few half-baked questions, mails questionnaires to his constituents, gets a 10 per cent return, and pretends to believe the conclusions, the best potential excuse is lack of both common sense and elementary knowledge about research--that is, a charitable assumption that he believes what he is getting. The Cabinet minister, confirming his gut feelings by simplistic polls, may not be much better than leaders of an earlier day who relied on their astrologers. But at least he is trying. And he will undoubtedly continue to try. When his staff bring him better channels he will use--and occasionally abuse--them. Such channels exist, not only for politicians, but for the public service as well.

Certainly, if obtaining a better understanding of the views of citizens is the chief concern, incasting cannot be the main channel, but it may be useful for bringing in surface-deep information about opinions that the respondent is aware of and is able to identify, and which can be extracted by analysts from two to five other possible answers. It cannot go deeper.

It may serve, as well, to register a response on a simple broadcasted message, limited in length and depth. In this case it will represent more of a dialogue than would occur from watching television, but not much more. Finally, the technique can be used to work quickly, after a programme has been broadcast, but the method is by no means foolproof. It cannot guarantee freedom from pressure or assure that the apparent respondent has in fact voted. On the other hand, once a vote has been recorded there are not sufficient guarantees of the anonymity of the respondent.

It should also be recognized that electronic social incasting will be only slowly adapted to the policy process. Broadcasts on the same topic cannot be made too frequently. Moreover, one can imagine that the content will be influenced by many compromises, mainly by the business concerns of the broadcasting network. Even so, there may be an improvement over the *status quo*, if we are right in our hunch that social incasting may have an impact on the motivation of participants.

Viewers will likely become more interested in social issues, being amused at least in the beginning by the possibility of feedback. Later, they will react to the way the feedback is used: by an effort to participate more, by frustration, or by a turning away all together. If the reactions are negative, they may in fact contribute to an increased pressure for improvement of the process of government-citizen communication.

Citizens' groups, pressure groups, and their advocates will probably share the hopes and disillusionments of the population as a whole. Increased popular interest in the responsiveness of governments may perhaps improve their position, which in turn may increase their motivation.

In summary, then, governments may well be confronted by popular pressure to improve communication with the public. Each government and within it perhaps each agency, will react according to its own specific situation at a given time: either to use the opportunity to introduce mechanisms for better understanding of citizens or to fight back by reacting passively or defending the *status quo*. Forcing this choice may be the most relevant impact of social incasting; obviously, this impact will be only limited.

COMPLICATIONS

In the face of the possible introduction of social incasting, some unique features of the Canadian situation should be taken into account:

a. The existing federal-provincial conflict over responsibility for communications will make control of incasting more difficult; also the conflict itself may be exacerbated by abuses of incasting.

One example of potential conflict arises from the hearings by the Alberta Public Utilities Board which started in August 1979. It is assumed that technology for incasting and its impact fall under the jurisdiction of the board. There is no certainty, however, whether the board should exercise its jurisdiction, given the regulating activities of the Canadian Radio-television and Telecommunications Commission. This commission has already given a ruling in one case (in Ontario), but has not yet accepted any explicit policy for incasting.

b. Social communication in Canada is hampered by conditions that are unique to industrially developed countries: the unusual distribution of population (90 per cent in a strip with a width/length ratio of 1:12, the rest with an average density under 0.2 persons/km^2); the existence and legitimation of two major and several minor cultures; and a relatively short tradition of nation-wide concerns and contacts. Any improvement or deterioration in peoples' motivation to communicate can hit a socially sensitive point. Incasting could be one of such factors.

c. The traditionally strong position of the public service in the policy-making process makes improvements in communication *to* governments different from several other countries: more necessary, more difficult to start and, hopefully, more co-ordinated once started.

d. The United States, as the single, strong neighbour, has a heavy influence in matters of life-styles and behaviour patterns (including the potential popularity of incasting) and as an economic partner and competitor (potential exporter of hardware and programmes). This makes it difficult both to withstand outside pressure and, because there are social differences between the two countries, to learn from our neighbour's experiences.

One issue where influences from the United States could be very important is related to the control of the system--legally, economically, and ideologically. There are strong pressures in the United States to move the jurisdiction for communications from the federal level. In Canada, this same issue involves the whole communication field, including some specific questions such as pay television. The main driving forces here are the federal-provincial rivalry for power and the need to find an optimum balance between nation-wide and provincial needs in questions of culture, education, and economic and political impact. In the United States, however, the driving force in the anti-federal campaign seems to be business interests. Pushing for a *Consumers Communication Reform Act*, they hope that a transfer of power from the Federal Communications Commission to the states would weaken the government control and strengthen their own position.

e. Finally, it is necessary to mention the conflict that exists between the carriers, responsible for the transmission, and the operators, responsible for the content. The advent of the cable television industry has triggered questions about delimitation and the optimal degree of concentration of companies in the field. Problems relating to the possible introduction of electronic social incasting, especially those related to investment and regulation, will have to be solved in a milieu of considerable flux.

CONCLUSIONS

A general guess about the future impact of social incasting is, as we have seen, uncertain, but some selected conclusions emerge that seem probable enough to be a basis for potential action. They are presented here as they have been synthesized: as a set of explicit numbered statements, each to be judged on its own merit, but all of them mutually interdependent.

1. *Technology of Incasting*

It is nearly certain that the technology has already been developed, or will be developed in the near future, to a point where mass application is possible.

2. *Economic Impact of Incasting in Canada*

If introduced, incasting could have a considerable economic impact on the television business, especially via

- Its entertainment value
- Its utility for evaluating audiences
- Its evaluation of commercials and, through it, perhaps its capacity to increase the pressure to buy
- Sharing technology with pay television (for technologies using cable television)
- Education.

If incasting were introduced, its indirect economic impact would be considerably greater than its share of the direct impact from manufacturing the related technology. (Direct impact corresponds to sales volumes either of specialized incasting hardware or of a share in the videotex business, depending on the choice of the technology used.) For economic reasons, the introduction of incasting in Canada is, in general, probable; for cable television it is highly probable.

3. *Influence of Government Economic Assistance, Especially Funding, on the Introduction of Incasting in Canada*

Federal help can be instrumental in the technological development of the product and the introduction of its results. Without it, foreign technology would have to be used and almost certainly more hardware would be imported. This would mean a decrease of employment, profits, and the opportunity to learn for Canadians. Withdrawal of government help could also delay the timing of its introduction. *It could not*, however, *prevent or considerably limit its introduction.* The indirect impact is strong enough to cause imports of the equipment if the need is perceived to be there.

4. *Other Possibilities to Influence the Introduction of Incasting to Canada*

Let us suppose that incasting is found to be technologically and economically feasible, but that Canadians would wish to influence its introduction to Canada. If no specific action is taken, incasting will nevertheless be introduced for the variety of reasons mentioned above. To prevent its introduction, only strong government action could be effective. Such action, however, seems improbable and even impossible. Because of the indirect impacts the pressure for introduction will be strong. If Canadian technology is used and/or Canadian industry participates, the expected direct impact will cause even greater pressure. It is highly improbable that the arguments against introduction (most of them non-economic, long range, and speculative) will generate enough counter-pressure. Unless an extremely strong government, unhampered by federal-provincial conflicts, were to make it a priority issue, no effective action can be expected. For all practical purposes, the decision on whether or not incasting should be introduced in Canada is out of the hands of any individual, group, or organization.

5. *Use of Incasting for Elections*

There is no way to identify the person voting and to guarantee his privacy and freedom from pressure. These are the main unsolved questions over anonymity. For this reason, incasting for elections is not possible. As the *Elections Act* in its present form is part of the image, reality, and myth of Canadian democracy, strong pressure leading to an abuse of incasting for election purposes seems highly improbable.

6. *Use of Incasting for a Referendum*

The same difficulties as in the case of elections make incasting inappropriate for a referendum. This view may not, however, always be shared by the whole Canadian population. Recent experiences with the insensitivity of governments have increased the feeling of frustration of the Canadian population, and are decreasing their confidence in the present process of social, especially government, decision making. The idea of direct democracy, with frequent referenda as a major tool, is becoming increasingly popular. Incasting, if discussed at all, is occasionally welcomed as a technological fix even by persons who proclaim themselves enemies of the technological fix. Direct democracy in this context is both a programme and an emotional proclamation of anger, alienation, and distrust. As such it is not always open to arguments about the limits on the role of referenda. If this trend continues some results of incasting may be interpreted as a referendum, rather than as informative communication, and pressure might be brought to bear on decision makers to act accordingly. Compared to the risk of misinterpretation of social incasting generally, the danger of this occurring is not very significant. The probability of *open abuse* of incasting for conducting referenda is still lower. Nevertheless, the existence of incasting may increase the pressure for direct democracy which, in turn, could increase the pressure for referenda by traditional means.

7. *Social Incasting as a Contribution to Government Responsiveness*

The process by which the government understands the attitudes and perceived needs of citizens is a major bottle-neck in the function of many Western countries, including Canada. If it shows improvement, interactive broadcasting will be one component of it, contributing in the following way:

a. The citizens will feel that they are being heard. This will decrease their frustration a little and motivate them to make more effort to participate through incasting and other channels.

b. Interaction increases the popularity of a programme and the span of the viewer's attention. As a result, the education of citizens via the media will improve to a degree.

c. Incasting will contribute to the upstream information necessary for citizens' understanding. This information will be fast, inexpensive, based on medium-intensive interactive communication, not very reliable, and limited in length and depth. Integrated with other channels it will play its own limited, but not negligible, role.

8. *Impact of Social Incasting if Governments Remain Unresponsive*

Participants will feel that their opinion has no impact and is not being heard by decision makers. After the initial euphoria, this may contribute to a deepening frustration and reorientation of citizens from informative communication to other avenues of potential (or imaginary) power. Lack of overall government responsiveness will probably have the same effect concurrently and the impact of incasting in this direction will be only limited.

Other negative impacts have to be expected. Some persons may feel that incasting has solved the problem of communication between the government and citizens and slow down in other efforts to improve it. For others, incasting may at the beginning strengthen the mistaken belief that human communication can be solved by technology, by simplistic processes ("just push the button"), or by the computation of numbers. Further, unless specific regulations for interactive broadcasting are controlled quickly, there may be occasional breaches of anonymity.

It is quite probable that if the process of communication continues to degenerate it would later lead, through increased conflict, to an increased responsiveness of governments anyway. Interactive broadcasting may play, in this case, the typical role of a potential partial social improvement: if neglected in its own time, it marginally contributes to an increase in conflict and, through it, to a more pervasive change later. Only someone sufficiently detached to regard this bifurcation as merely two different routes to improvement can evaluate interactive broadcasting as a tool for certain improvement.

9. *Impact on Local Government*

Compared with the impact on senior levels, incasting can probably be used on the local level with less concern for co-ordination with other government processes. It can contribute to "feeling of community" and increase the sensitivity of local decision makers. Probably it can even prevent extremely unpopular actions, occasionally including actions harmful to the community.

10. *Other Impacts of Social Incasting on Citizens at Large*

An increase in the motivation to participate (nearly certain at the beginning and possible permanently) will increase the information of citizens about socio-political issues. The results of incasting will contribute, to a lesser or greater degree, to the direct understanding of citizens by citizens. In this sense, the impact will be similar to the impact on communication with governments, but less complicated.

11. *Impact on Broadcasting*

Social incasting may make broadcasting organizations better and more quickly aware of the preferences of their viewers and hearers. In an effort to improve ratings, this may lead to an immediate effort to cater to every whim of customers and decrease the influence of the professional staff. (There is some criticism that many mass media are already "run by polls" anyway.) Perhaps it could lead to a renewed discussion among broadcasters and even to some tentative tactical conclusions about the role of the professional staff in directing the media.

Many related questions will need further study before even tentative answers can be found. To be further explored are the impact on mass programming and targeted programming; the potential dilution of the market (with the concurrent impact of transmission via satellites); the relations between carriers and operators; and the relation of both of these to the print media, and so on.

Evidently, incasting will change both the motivation of participants and the content of the media as described above.

12. *Danger of Manipulation of the Public*

Electronic incasting channels will certainly increase the opportunity to manipulate and the effort to do so. In the socio-political domain, success will depend on the situation in the political power game. Nevertheless, it does seem unlikely that in the next decade any one organized group in Canada will enjoy sufficient power to manipulate the public successfully.

According to some opinions, there is on the horizon a period of increased concern with human values and ideologies, the "no-more-ideologies" thesis being regarded as one of the ideologies. The present state of discussion of this *problématique* is not sufficiently developed to serve as a basis for a forecast related to incasting. Theoretically, however, it could offer a good domain for manipulation.

The situation is different with respect to commercials and consumer information. Increased manipulation of consumer demand and taste is very probable and its impact on our life-style quite possible. The methodology for manipulating citizens will probably improve and we shall, on average, buy as told. Rather than being "automated citizens," we shall, in this respect at least, perhaps be "50 per cent automated citizens."

13. *Possibility to Control Social Incasting by Regulations Specific to Incasting*

If incasting is introduced in Canada, its applicability for political purposes will be obvious from the beginning. In the absence of regulations, incasting will become a weapon in the political power game. Its advent will be considerably faster than has been the case with other political weapons and, as a consequence, its "rules of the game" more rudimentary, less functional, and more permissive. Because of the complexity surrounding the issue, work on the regulations will not be done quickly. If the beginning of the work has to await strong public pressure, the results will be ready at a time when the fight will already be in full swing. It is improbable that in this case, under any foreseeable circumstances, the warring parties will allow the introduction of effective regulation in Canada.

Broadcasting professionals are traditionally ard increasingly suspicious of governments' intentions concerning regulation in the broadcasting domain. If governments try to solve interactive broadcasting in isolation from other issues, opposition can be expected. Its strength can only be a matter of conjecture. If, on the other hand, broadcasters see an opportunity to help with the problem of making governments more responsive, their attitude to the preparation of regulations might change. At present, the Canadian economy is generally perceived as being overregulated. In an atmosphere of deregulation it may not be easy to show that incasting needs regulation, that it will get it anyway, and that delays will only increase the level of conflict around it. Even if the work on regulations begins reasonably soon, their introduction will be a matter of political judgement and will be very sensitive to pressures. Without a change in the overall process of understanding citizens by goverrments, however, no regulation in this domain can be more than a palliative.

GENERAL CONCLUSION

Business interests, without caring much about social incasting, will probably soon introduce to Canada technologies and organizations that will make social incasting possible. Once introduced, social incasting will contribute to the pressure for improvement of overall government-citizen communication. If governments undertake a profound revision of this process of communication, incasting may marginally help. If they do not, incasting will marginally exacerbate the coming conflict.

RECOMMENDATIONS

The conclusions, as summarized, can lead to several sets of recommendations. Three selected recommendations, addressed to the Government of Canada, conclude the present text.

THE PRINCIPAL RECOMMENDATION

To start, as soon and as intensively as possible, to make improvements in the process by which the government (a) understands the perceived needs and policy-related attitudes of Canadians; and (b) anticipates the reactions of Canadians to government actions. In this context, which is crucial both for the incasting and for much broader concerns, to identify and facilitate the main role of interactive broadcasting.

THE SECOND RECOMMENDATION

To prepare regulations for interactive broadcasting while not overestimating how much any regulation can influence interactive broadcasting in Canada. It is desirable to start the work soon and continue as much as possible without pressure, to try to achieve at least mutual communication between actors, before the positions become petrified and the conflict starts in full.

THE THIRD RECOMMENDATION

To spend government funds on the development of Canadian interactive broadcasting according to economic criteria (for example, employment, profits, learning process), but to avoid the belief that government funding can crucially influence the existence of interactive broadcasting in Canada (i.e., the question of whether or not it will be introduced).

NOTES

1. For a discussion of the use of the home computer for performing these and other functions see C.C. Gotlieb, *Computers in the Home: What They Can Do for Us--And To Us* (Montreal: Institute for Research on Public Policy, 1978).

2. Other comparable, but quite advanced, systems have been introduced abroad. The British Post Office developed Prestel (known as Viewdata at the beginning) and started market trials in June 1978 with 80,000 pages of information and 1,500 users. Public use in 1979 was expected to start with 200,000 pages of information. Germany and Holland have already purchased Prestel, but two other systems are also being tested in the United Kingdom: Ceefax and Oracle, developed by the British Broadcasting Corporation and the Independent Broadcasting Authority. France is working on her Titan system and as an alternative is discussing licensing arrangements to import Canadian videotex. This year, worldwide technical standards will probably be set. If they are based on Telidon, which is quite possible, it would be a help to Canada. The United States and Sweden have not yet committed themselves, and Britain is aggressively selling Prestel, which is cheaper.

3. In a communication, Gordon Thompson has stressed the differences between videotex and incasting: "In a videotex system, the system must be capable of generating within the population authors who will supply the content to keep it alive. My studies indicate that because of the *unique* potential of the videotex system to allow its content to behave as a private economic good, there are really no decent examples of content flourishing in the market today. Information, in the past, has been largely treated as a public economic good, because of its close and intimate association with the material good used to carry the information...be it a book, a record or whatever. Without a vital and economically viable content the whole notion of videotex whithers away. The transactions involved in the content must be more significant by an order of magnitude perhaps, than those involved in the acquisition and maintenance of the hardware. To paraphrase McLuhan, the content must be the economic message of videotex services. Because of this critical and usually overlooked important aspect of content, videotex becomes very different than the more simplistic things like Qube, pay-TV and various other "interactive TV." For example, videotex is totally different from teletext in this regard, for teletext forces its content to forever behave as a public economic good unless very heavy externalities are absorbed. Although a videotex system could perform some incasting type chores, it would be, in my opinion, a gross misuse of the system." For a fuller discussion of Gordon Thompson's views see his study *Message from Mercury: Information Technology* Is *Different* (Montreal: Institute for Research on Public Policy, 1979).

4. Twenty-three interesting papers have been published in the Special Issue on Social Implications of Telecommunications, *IEEE Transactions on Communication*, Volume CCM-23, No. 10, October 1975. Several of them are directly or

indirectly related to incasting. Kenneth C. Laudon, in his *Communications Technology and Democratic Pariticipation* (New York: Praeger, 1977), has a useful bibliography.

Institute for Research on Public Policy

PUBLICATIONS AVAILABLE*

July 1980

BOOKS

Leroy O. Stone & Claude Marceau	*Canadian Population Trends and Public Policy Through the 1980s.* 1977 $4.00
Raymond Breton	*The Canadian Condition: A Guide to Research in Public Policy.* 1977 $2.95
Raymond Breton	*Une orientation de la recherche politique dans le contexte canadien.* 1978 $2.95
J.W. Rowley & W.T. Stanbury, eds.	*Competition Policy in Canada: Stage II, Bill C-13.* 1978 $12.95
C.F. Smart & W.T. Stanbury, eds.	*Studies on Crisis Management.* 1978 $9.95
W.T. Stanbury, ed.	*Studies on Regulation in Canada.* 1978 $9.95
Michael Hudson	*Canada in the New Monetary Order: Borrow? Devalue? Restructure!* 1978 $6.95
W.A.W. Neilson & J.C. MacPherson, eds.	*The Legislative Process in Canada: The Need for Reform.* 1978 $12.95
David K. Foot, ed.	*Public Employment and Compensation in Canada: Myths and Realities.* 1978 $10.95
W.E. Cundiff & Mado Reid, eds.	*Issues in Canadian/U.S. Transborder Computer Data Flows.* 1979 $6.50
G.B. Reschenthaler & B. Roberts, eds.	*Perspectives on Canadian Airline Regulation.* 1979 $13.50
P.K. Gorecki & W.T. Stanbury, eds.	*Perspectives on the Royal Commission on Corporate Concentration.* 1979 $15.95
David K. Foot	*Public Employment in Canada: Statistical Series.* 1979 $15.00

Order Address: The Institute for Research on Public Policy
P.O. Box 9300, Station "A"
Toronto, Ontario
M5W 2C7

Meyer W. Bucovetsky, ed.	*Studies on Public Employment and Compensation in Canada.* 1979 $14.95
Richard French & André Béliveau	*The RCMP and the Management of National Security.* 1979 $6.95
Richard French & André Béliveau	*La GRC et la gestion de la sécurité nationale.* 1979 $6.95
Leroy O. Stone & Michael J. MacLean	*Future Income Prospects for Canada's Senior Citizens.* 1979 $7.95
Douglas G. Hartle	*Public Policy Decision Making and Regulation.* 1979 $12.95
Richard Bird (in collaboration with Bucovetsky & Foot)	*The Growth of Public Employment in Canada.* 1979 $12.95
G. Bruce Doern & Allan M. Maslove, eds.	*The Public Evaluation of Government Spending.* 1979 $10.95
Richard Price, ed.	*The Spirit of the Alberta Indian Treaties.* 1979 $8.95
Peter N. Nemetz, ed.	*Energy Policy: The Global Challenge.* 1979 $16.95
Richard J. Schultz	*Federalism and the Regulatory Process.* 1979 $1.50
Richard J. Schultz	*Le fédéralisme et le processus de réglementation.* 1979 $1.50
Lionel D. Feldman & Katherine A. Graham	*Bargaining for Cities: Municipalities and Intergovernmental Relations, An Assessment.* 1979 $10.95
Elliot J. Feldman & Neil Nevitte, eds.	*The Future of North America: Canada, the United States, and Quebec Nationalism.* 1979 $7.95
Maximo Halty-Carrere	*Technological Development Strategies for Developing Countries.* 1979 $12.95
G.B. Reschenthaler	*Occupational Health and Safety in Canada: The Economics and Three Case Studies.* 1979 $5.00
David R. Protheroe	*Imports and Politics: Trade Decision-Making in Canada.* 1980 $8.95
G. Bruce Doern	*Government Intervention in the Canadian Nuclear Industry.* 1980 $8.95

G. Bruce Doern & R.W. Morrison, eds.	*Canadian Nuclear Policies.* 1980 $14.95
W.T. Stanbury, ed.	*Government Regulation: Scope, Growth, Process.* 1980 $10.95
Yoshi Tsurumi with Rebecca R. Tsurumi	*Sogoshosha: Engines of Export-Based Growth.* 1980 $8.95
Allan M. Maslove & Gene Swimmer	*Wage Controls in Canada, 1975-78. A Study in Public Decision Making.* 1980 $11.95
T. Gregory Kane	*Consumers and the Regulators: Intervention in the Federal Regulatory Process.* 1980 $10.95
Albert Breton & Anthony Scott	*The Design of Federations.* 1980 $6.95
A.R. Bailey & D.G. Hull	*The Way Out: A More Revenue-Dependent Public Sector and How It Might Revitalize the Process of Governing.* 1980 $6.95
Réjean Lachapelle & Jacques Henripin	*La situation démolinguistique au Canada: évolution passée et prospective.* 1980 $24.95
Raymond Breton, Jeffrey G. Reitz & Victor F. Valentine	*Cultural Boundaries and the Cohesion of Canada.* 1980 $18.95

OCCASIONAL PAPERS

W.E. Cundiff (No. 1)	*Nodule Shock? Seabed Mining and the Future of the Canadian Nickel Industry.* 1978 $3.00
IRPP/Brookings (No. 2)	*Conference on Canadian-U.S. Economic Relations.* 1978 $3.00
Robert A. Russel (No. 3)	*The Electronic Briefcase: The Office of the Future.* 1978 $3.00
C.C. Gotlieb (No. 4)	*Computers in the Home: What They Can Do for Us--And to Us.* 1978 $3.00
Raymond Breton & Gail Grant Akian (No. 5)	*Urban Institutions and the People of Indian Ancestry.* 1978 $3.00
K.A. Hay (No. 6)	*Friends or Acquaintances? Canada as a Resource Supplier to the Japanese Economy.* 1979 $3.00

T. Atkinson (No. 7)	*Trends in Life Satisfaction.* 1979 $3.00
M. McLean (No. 8)	*The Impact of the Microelectronics Industry on the Structure of the Canadian Economy.* 1979 $3.00
Fred Thompson & W.T. Stanbury (No. 9)	*The Political Economy of Interest Groups in the Legislative Process in Canada.* 1979 $3.00
Gordon B. Thompson (No. 10)	*Memo from Mercury: Information Technology* Is *Different.* 1979 $3.00
Pierre Sormany (No. 11)	*Les micro-esclaves: vers une bio-industrie canadienne.* 1979 $3.00
K. Hartley, P.N. Nemetz, S. Schwartz, D. Uyeno, I. Vertinsky & J. Young (No. 12)	*Energy R & D Decision Making for Canada.* 1979 $3.00
David Hoffman & Zavis P. Zeman, eds. (No. 13)	*The Dynamics of the Technological Leadership of the World.* 1980 $3.00
Russell Wilkins (No. 13*a*)	*Health Status in Canada, 1926-1976.* 1980 $3.00
Russell Wilkins (No. 13*b*)	*L'état de santé au Canada, 1926-1976.* 1980 $3.00
P. Pergler (No. 14)	*The Automated Citizen: Social and Political Impact of Interactive Broadcasting.* 1980 $4.95
Zavis P. Zeman (No. 15)	*Men with the Yen.* 1980 $5.95

WORKING PAPERS (No. Charge)**

W.E. Cundiff (No. 1)	*Issues in Canada/U.S. Transborder Computer Data Flows.* 1978 (Out of print; in IRPP book of same title.)

Order Working Papers from
The Institute for Research on Public Policy
P.O. Box 3670, Halifax South
Halifax, Nova Scotia
B3J 3K6

John Cornwall (No. 2)	*Industrial Investment and Canadian Economic Growth: Some Scenarics for the Eighties.* 1978
Russell Wilkins (No. 3)	*L'espérance de vie par quartier à Montréal, 1976: un indicateur social pour la planification.* 1979
F.J. Fletcher & R.J. Drummond (No. 4)	*Canadian Attitude Trends, 1960-1978.* 1979